REARING YOUNG STOCK ON TROPICAL DAIRY FARMS IN ASIA

John Moran

PUBLISHING

National Library of Australia Cataloguing-in-Publication entry

Moran, John, 1945–

Rearing young stock on tropical dairy farms in Asia/by John

Moran.

9780643107427 (pbk.)
9780643107915 (epdf)
9780643107922 (epub)

Includes bibliographical references and index.

Dairy cattle – Asia.
Calves – Asia – Growth.
Heifers – Asia.
Dairy farming – Asia.

636.2142095

Published by
CSIRO PUBLISHING
36 Gardiner Road, Clayton VIC 3168
Private Bag 10, Clayton South VIC 3169
Australia

Telephone: [+613] 9545 8555
Local call: 1300 788 000 (Australia only)
Fax: +61 3 9662 7555
Email: csiropublishing@csiro.au
Web site: www.publishing.csiro.au

All photographs are by John Moran unless otherwise indicated

Set in 10.5/13 Adobe Minion Pro and Helvetica Neue LT Std
Edited by Peter Storer Editorial Services
Cover and text design by James Kelly
Typeset by Desktop Concepts Pty Ltd, Melbourne
Printed by Ingram Lightning Source

Mar26_RP_ILS

Foreword

It gives me great honour and pleasure to write this Foreword for the book *Rearing Young Stock on Tropical Dairy Farms in Asia*, authored by Dr John Moran from Victoria in Australia.

Dairy farmers on small holder farms in the tropics are continually being faced with decreasing profit margins, arising from increasing costs of milk production and milk prices that do not reflect these costs. They should always be reassessing their farm practices to determine which farm costs can be reduced and which aspect of their farm management can be improved to increase milk yields hence on-farm cash flows.

Rearing young stock is probably one of the least well understood and practiced aspects of herd and feeding management on dairy farms throughout the world, particularly on small holder dairy farms in the tropics, for a variety of reasons:

- Replacement heifers do not generate any income for 2 or 3 years.
- Mortality rates of milk-fed and weaned heifer replacements in the tropics are frequently double those on temperate farms.
- Poorly reared heifers are more susceptible to diseases, so frequently have a shorter lifespan in the milking herd.
- Small heifers are less productive both in terms of daily milk yields and the number of calves produced during their lifetime.

To overcome these problems, rearing replacement heifers must be considered as an investment in the farm's future rather than a farm cost.

For many years, Dr John Moran has been considered as an expert in tropical dairy farm production, particularly in South and East Asia. Over the last 10 years, he has conducted over 90 workshops and written several books specifically directed to the tropical small holder dairy farmer. John wrote his first calf- and heifer-rearing book 20 years ago and this book is an updated version but with the emphasis on the specific problems dairy farmers must address when rearing their replacement stock in the harsh tropical climate. Together with the valuable sources of information he cites in this manual, this book also serves as a training manual.

Not only will this book be of great interest to dairy farmers and advisers throughout the tropics, but university and college students, dairy scientists and extension personnel will also find it a valuable reference text on young stock management. I therefore highly recommend this book to all.

Professor Dr Metha Wanapat
Director, Tropical Feed Resources Research and Development Centre (TROFREC)
Faculty of Agriculture, Khon Kaen University
Khon Kaen, Thailand
Email: metha@kku.ac.th

About the author

For 30 years, Dr John Moran was an Australian senior research and advisory scientist from Victoria's Department of Primary Industries (DPIV), located in northern Victoria. Over the last 10 years, he spent half his time advising farmers in southern Australia and half his time working with dairy farmers and advisers in South and East Asia. His specialist fields include dairy production, ruminant nutrition, calf and heifer rearing, forage conservation and whole farm business management. Following his retirement from DPIV in Jun 2011, John formed a consulting business, Profitable Dairy Systems, located in Kyabram, Victoria.

John graduated in 1967 with a Rural Science honours degree from New England University at Armidale in NSW, followed by a Masters degree in 1969. In 1976, he obtained a Doctorate of Philosophy in beef production from the University of London, Wye College in England. During the 1980s, John lived in Indonesia for 3 years, working in beef cattle and buffalo research. Since 1999, he has initiated and conducted training programs on small holder dairy production to farmers, advisers and policymakers in Indonesia, Malaysia, Thailand, Vietnam, China, Pakistan, Sri Lanka and East Timor.

As a result of his Asian programs, in 2005 John wrote his first dairy manual, *Tropical Dairy Farming: Feeding Management for Small Holder Dairy Farmers in the Humid Tropics*. As well as hard copy, the book was published on the internet, making it freely available to dairy stakeholders throughout the world. In 2008 John wrote his second manual on tropical dairy farming, *Business Management for Tropical Dairy Farmers*. By July 2012, the two books had received over 93 000 'hits' on the internet, indicating their relevance particularly to Asian tertiary teaching institutes and government livestock departments.

During 2009, John was commissioned by LIVECORP – Australia's coordinating agency for exporting cattle and sheep to many countries throughout the world – to develop a series of management packages for tropical dairy production technology. The background review for these packages form the basis of his recently published third book, entitled *Managing High Grade Dairy Cows in the Tropics*. This, his fourth book on tropical dairy farming, concentrates on rearing young stock – a topic on which John has published widely over the last 20 years.

John has published more than 200 research papers and advisory articles. He has also written several manuals on dairy and beef cattle nutrition, veal production, calf and heifer rearing and on silage production. The first edition of *Calf Rearing: A Guide to Rearing Calves in Australia*, published in 1993, sold more than 10 000 copies. The second edition, published in 2002, is still selling widely throughout Australia. He also published a companion book on young stock management: *Heifer Rearing: A Guide to Rearing Dairy Replacement Heifers in Australia.* His book, *Forage Conservation: Making Quality Silage and Hay in Australia*, published in 1996, is now a set text for undergraduate study in several Australian universities. One of his latest books, *Feedpads for Grazing Dairy Cows*, which he wrote in 2010 with his DPI colleague Scott McDonald, is also a best seller in the dairy industry. All books are commercially available through CSIRO Publishing on <http://www.publish/csiro/au>.

John's initial training in a systems approach to livestock science, together with his many years working closely with dairy industries in Australia and South and East Asia stands him in great stead to write this fourth tropical manual, which complement his previous books. *Rearing Young Stock on Tropical Dairy Farms in Asia* is both a training manual for farmers, dairy advisers, technical staff and other tropical dairy specialists plus a series of observations on how successful and profitable dairy farmers manage their young stock.

Other books and technical manuals by the author

Books

Calf Rearing: A Guide to Rearing Calves in Australia (1993)
Forage Conservation: Making Quality Silage and Hay in Australia (1996)
Heifer Rearing: A Guide to Rearing Dairy Replacement Heifers in Australia (with Douglas McLean) (2001)
Calf Rearing: A Practical Guide (2002)
Tropical Dairy Farming: Feeding Management for Small Holder Dairy Farmers in the Humid Tropics (2005)
Business Management for Tropical Dairy Farmers (2009)
Feedpads for Grazing Dairy Cows (with Scott McDonald) (2010)
Fifty Years of Farmer Extension for Victoria's Dairy Industry (2011)
Rearing Young Ruminants on Milk Replacers and Starter Feeds (with Uppoor Krishnamoorthy) (2011)
Managing High Grade Dairy Cows in the Tropics (2012)

Technical manuals

Maize for Fodder: A Guide to Growing, Conserving and Feeding Irrigated Maize in Northern Victoria (with Ken Pritchard) (1987)
Growing Calves for Pink Veal: A Guide to Rearing, Feeding and Managing Calves for Pink Veal in Australia (1990)
Growing Quality Forages for Small Holder Dairy Farms in Indonesia (2001)
Managing Dairy Farm Costs: Strategies for Dairy Farmers in Irrigated Northern Victoria (2002)
Feeding Management for Small Holder Dairy Farmers in Thailand (2002)
Improving Milk Composition Through Better Feeding Management (2003)
The Key Drivers of Good Reproductive Performance on Indonesian Dairy Farms (2006)
Value Adding Indonesia's Dairy Industry: Developing Cottage Industries in East Java (2006)

Improving Business Skills of Small Holder Dairy Farmers in Thailand (2007)
Dairy Production in Malaysia with Particular Reference to Milk Quality (2007)
Dairy Production in Indonesia with Particular Reference to Milk Quality (2007)
Developing a Post-Arrival Herd Management Protocol for Imported Australian Dairy Heifers (2007)
Managing Heat Stress and Housing for Pakistani Dairy Cows and Buffaloes (2007)
Dairy Production from Smallholder Farmers in Asia – Synopses of Workshops from 2000 to 2007 (2008)
Dairy Production in Sri Lanka: Current Situation and Future Prospects (2008)
Improving Business Skills in Vietnam's Small Holder Dairy Industry (2008)
A Guide to Better Dairy Herd Management in the Tropics (2009)
Best Management Practices for Small Holder Dairy Farmers in Asia (2010)
A Blue Print for a 50 Cow Farm in Asia (2010)
A Blue Print for a Large Scale Intensive Dairy Farm in Asia (2012)
Guide to e-modules on 'A Short Course in Tropical Dairy Farming' (2012)
Guide to e-modules on 'Feeding management on small holder dairy farms in the tropics' (2012)

Acknowledgements

Throughout the humid tropics, small holder dairy farming was established as part of social welfare and rural development schemes, to provide a regular cash flow for poorly resourced, and often landless, farmers. Now it is an accepted rural industry and requires a more business-minded approach to farm management. One such method of increasing the cash flow of small holder dairy farmers is to improve the feeding and herd management of their livestock through rearing better quality young stock. This is the goal of this technical manual.

The manual is based on one I wrote in 2002 for Australia's dairy industry. It has been extensively modified to ensure its relevance to small holder farmers in the humid tropics. I would like to acknowledge my colleagues in the Dairy Extension Centre of Victoria's Department of Primary Industry for their discussions and advice over many years in the fields of calf and heifer management.

Together with a team of Australian dairy specialists, I developed training programs for small holder dairy farmers throughout Asia on many aspects of small holder farm management, including young stock. I would like to acknowledge my colleagues:

- Frank Mickan (Victorian Department of Primary Industries, Ellinbank, Vic)
- John Miller (Queensland Department of Primary Industries, Murgon, Qld)
- Denise Burrell (National Centre for Dairy Education Australia, Terang, Vic)
- Jo Crosby (National Centre for Dairy Education Australia, Terang, Vic)
- Bill Tranter (Tableland Veterinary Service, Malanda, Qld)

Following the successful publication of my book *Tropical Dairy Farming* in December 2005, in 2009 CSIRO Publishing asked me to write a second manual, *Business Management for Tropical Dairy Farmers*. My two latest Asian dairy farming books (*Managing High Grade Dairy Cows in the Tropics* and this one, *Rearing Young Stock on Tropical Dairy Farms in Asia*) have also been published by CSIRO. As with my first two books, my gratitude also goes out to ATSE Crawford Fund, who provided generous financial support for all four books, and to Ted Hamilton and the staff at CSIRO Publishing who very professionally converted my laptop computer writings to both a 'hard copy' book and a collection of PDF files for an eBook.

Dr John Moran
Director, Profitable Dairy Systems,
24 Wilson St, Kyabram, Victoria 3620, Australia
Telephone: +61 418 379 652 (mobile phone)
Email: jbm95@hotmail.com

Acknowledgement of The Crawford Fund

The publication of this book would not have been possible without the generous assistance of The Crawford Fund whose mission is to increase Australia's engagement in international agricultural research, development and education for the benefit of developing countries and Australia.

Chemical warning

The registration and directions for use of chemicals can change over time. Before using a chemical or following any chemical recommendations, the user should **always** check the uses prescribed on the label of the product to be used. If the product has not been recently produced, users should contact the place of purchase, or their local reseller, to check that the product and its uses are still registered. Users should note that the currently registered label should **always** be used.

Contents

1

Introduction

This chapter presents an outline of the book, which provides technical information on the theories, as well as the practices, of rearing young stock on tropical dairy farms.

The main points in this chapter

- This book is the fourth book written on various aspects of tropical dairy farming.
- Young stock (milk-fed calves and growing heifers) are the most neglected class of stock on most dairy farms.
- Of the nine key activities on dairy farms, young stock management generally receives the least attention. This is primarily because it has the greatest time span between any financial investment and return.
- Even on the best dairy farms, heifers will not generate income for 2 yr after they enter the dairy herd as newborn calves.
- If farm incomes decrease and cost savings become necessary, replacement heifers are all too frequently the first ones to suffer from reduced farm inputs.
- On farms where growing heifers can be moved away from the rest of the dairy herd, many problems associated with 'out of sight, out of mind' can occur.
- Improved management strategies leading to lower calving intervals, higher calving rates, reduced stillborn and pre-weaned calf mortalities and fewer non-pregnant heifers can supply many more dairy herd replacements.
- Such strategies can increase the number of replacement heifer calves in the herd from 15 to 36%, thus allowing farmers to increase their herd sizes naturally.
- This book highlights many of these problems, as well as those arising from the fact that heifers, particularly milk-fed calves, are the most susceptible class of stock to poor farm management practices.

This book is a companion to three previous books I have written on small holder dairy (SHD) farming in the tropics. The first book, *Tropical Dairy Farming* (Moran 2005), details the production technology of SHD farming, with the emphasis on nutrition and feeding management. The second book, *Business Management for Tropical Dairy Farmers* (Moran 2009), discusses the farm business management (FBM) skills required to ensure

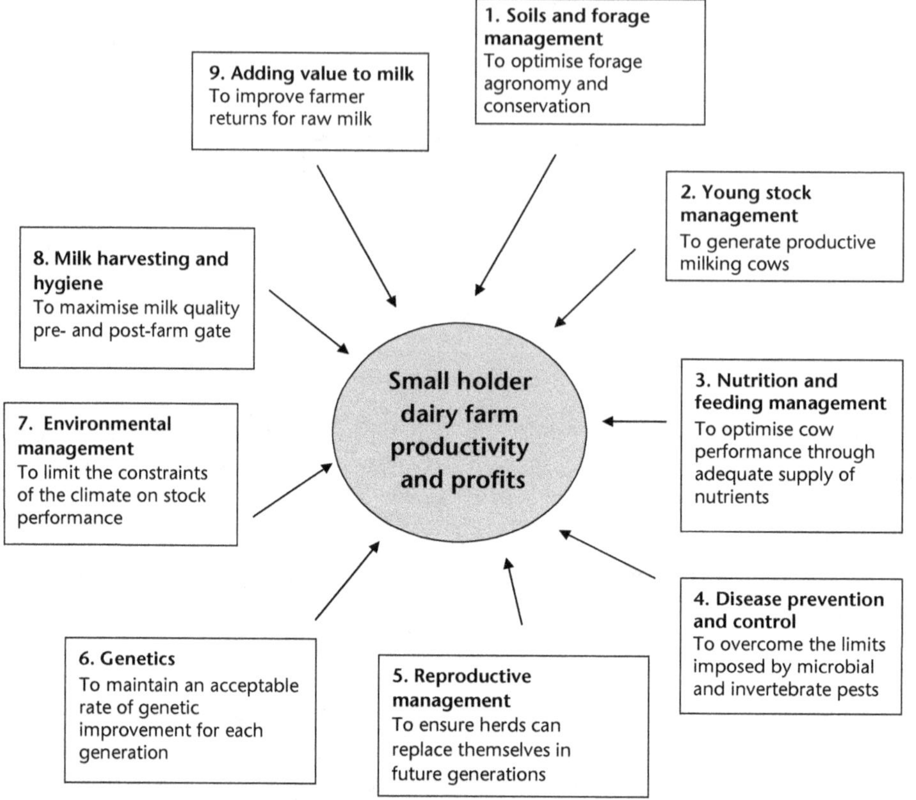

Figure 1.1. The nine links in the supply chain for a profitable dairy enterprise

such systems can remain financially sustainable. The third book, *Managing High Grade Dairy Cows in the Tropics* (Moran 2012), deals specifically with a major problem encountered by many tropical dairy farmers, namely the poor performance of exotic, high grade (that is high genetic merit) dairy cows when exported from their country of origin to a new, more stressful environment. This fourth book, *Rearing Young Stock on Tropical Dairy Farms in Asia*, deals with the most neglected of stock on most dairy farms, namely milk-fed and growing heifers.

1.1 The importance of young stock management on any dairy farm

Any financially viable dairy farmer, no matter where he or she is located, requires skills in a wide range of farm management practices, to be able to produce milk for considerably less money than consumers will have to pay for it once that milk enters into the supply chain of dairy processing, distribution and retail marketing.

Dairy farmers are fortunate in that, with raw milk in high demand, infrastructure is generally in place to sell their raw milk into the market place. Therefore their major efforts

A well-managed calf-rearing shed (Sri Lanka).

should be placed not in seeking markets, but on reducing its cost of production to provide an acceptable profit margin. This is the key objective of profitable dairy farming, whether the farmer manages five or 10 cows on a small holder traditional farm in the tropics, a larger operation with say, 200 cows predominantly grazing temperate pastures in the more established dairy industries of Australia or Europe, or a large-scale dairy feedlot located to strategically source forage crops and agro-industrial by-products anywhere in the world.

Figure 1.1 highlights the nine key dairy farm management practices that can be listed as links in the supply chain of a profitable dairy enterprise. The supply chain is the conversion of farm resources into saleable raw milk. The term 'enterprise' is used because many farmers, particularly those on small holder tropical farms, incorporate milk production into their mixed farming business. This figure details only those key activities in dairy production technology. It does not consider other equally important farmer activities that could be listed under the category of farm business management and, for larger farms, people skills. As with any commercial operation, the chain is only as strong as its weakest link, so each step needs to be properly managed. Weakening any one link through poor decision making can have severe ramifications on overall farm performance, and hence profitability.

Of these nine key links in profitable dairy farming, young stock management is the one that all too often receives the least attention. This occurs primarily because this link is the one with the greatest time span between financial investment and return. In other words, with the other eight links, the impact of decreasing farm inputs (which includes skills, time and money) will become more apparent as reduced farm returns. Conversely, increasing other farm inputs will return greater farm profits more quickly compared with improving young stock management.

One major concern of SHD farming throughout the tropics is the rearing of replacement heifers. The objective is to attain optimal growth so they can calve at an appropriate early age and at the lowest cost. They should promptly substitute for culled cows in the milking herd, so providing continuing returns on the investments of feed, labour and other farm resources. Success in raising replacement heifers is of major importance to the viability and profitability of any dairy enterprise. Rearing costs should then not be seen as herd costs, but as farm investments.

Even on the best dairy farms, heifers will not generate income for 2 yr after they enter the dairy herd as newborn calves. On many tropical farms, this can extend to 3 yr or even more. Consequently, any money spent on the young stock must be considered a relatively long-term investment. If farm incomes decrease and cost savings become necessary, replacement heifers are all too frequently the first ones to succumb to reduced farm inputs.

Furthermore, young stock can suffer through reduced non-cash farm inputs, particularly when they are weaned off milk and become the most neglected stock of the dairy herd enterprise. Milk-fed calves and milking cows require twice daily attention (feeding and milking respectively) while non-lactating stock (be they adult cows or growing heifers) can be, and often are, managed with less input. So long as they have access to drinking water, some forages and stay disease free, they will survive. On farms where replacement heifers can be moved away from the rest of the dairy herd, the situation can get even worse with the many problems associated with the 'out of sight, hence out of mind' scenario.

If SHD farmers cannot achieve realistic targets for heifer mortalities, live weight gain, mating age and age and live weight at first calving, they should look for a competent farmer who can, and then consider contracting that farmer to grow out their replacement stock. Another alternative would be to sell all their calves at birth or weaning and purchase older replacement heifers when required. However, purchasing all the replacements risks introducing disease onto the home farm and there is no guarantee that the genetic merit of the purchased stock is as good as the existing milking herd.

Low reproductive rates and high calf mortality are the major causes of reproductive wastage. This has a direct bearing on culling and replacement strategies and on genetic improvement. In many tropical countries, calf mortality can be as high as 50%. In some areas, this can be due to unacclimatised temperate or crossbred stock in addition to poor management. Climatic stress compounds the other hazards of calf life, but high calf mortality is usually due to diseases and poor feeding management.

This book highlights many of these issues, as well as those arising from the fact that heifers, particularly milk-fed calves, are the most susceptible of all dairy stock to poor farm management practices. This is due to their reduced resistance to the dairy farm environment (namely disease and climatic stress) and the fact that, until they become fully functioning ruminants, their monogastric digestive system requires a specific feed: namely milk.

Milk-rearing calves in Pakistan.

1.1.1 Benefits of bigger heifers

It is a well-researched fact (Moran and McLean 2001) that heavier heifers:

- get into calf easier the first time
- produce more milk in their first lactation
- get back in calf sooner during their first lactation
- stay in the herd longer
- need less help at calving
- cope better with herd competition
- put more money into the farmer's pocket.

Light heifers reduce herd fertility two ways: through delayed first calving and by having a longer period to get back in calf after first calving.

1.2 Outline of the manual

This book is written primarily for the stakeholders of SHD production in the tropics, although it is equally relevant to any tropical dairy farmer. Small holders are the major suppliers of milk in the tropics. However, numerous larger farms with many hundreds of milking cows, using intensive feedlot or less-intensive grazing systems, have been established throughout Asia in recent years, in an attempt to satisfy the increasing demand for fresh milk. However, I believe that, SHD farming is, and always will be, the backbone of domestic milk production throughout the tropics (Moran 2005, 2009, 2012).

Chapter 2 provides a concise summary of all the key elements of young stock management outlined in this book. Chapters 3, 4 and 5 review the key principles of nutrition of young stock: namely digestion of feeds in the milk-fed calf (Chapter 3), the nutrient requirements of calves (Chapter 4) and the importance of colostrum to newborn calves (Chapter 5). Chapter 6 reviews the recent published literature surveying calf and heifer mortalities in the tropics, while Chapter 7 discusses the facilities farmers require for rearing their young stock; it also discusses the important issue of ensuring good calf welfare.

Chapters 8–12 discuss the important components of the milk-rearing phase: namely feeding whole milk (Chapter 8), feeding calf milk replacer (Chapter 9), solid feeds (Chapter 10), disease management (Chapter 11) and communicating with the calves (Chapter 12). The next two chapters deal with feeding weaned replacement heifers (Chapter 13) and their management during mating and pre-calving (Chapter 14).

The remaining chapters of the book discuss improvements to the system of rearing replacement dairy heifers. These are the business aspects, or the costs and potential returns, from heifer rearing (Chapter 15), and assessing current rearing systems (Chapter 16), while Chapter 17 details some of the training programs that I have conducted around Asia in recent years. Finally, the best management practices for rearing young dairy stock are presented in Chapter 18.

Full publication details of all sources of information are presented in References. Every profession has its jargon, or words developed specifically for that profession, and agriculture is no exception. There are some very specific terms and acronyms that are routinely used by dairy researchers and advisers. These are explained in the Glossary and when they are first used in this book.

Appendix 1 lists what I have called 'John Moran's golden rules for calf and heifer management'. These were originally developed for Australian dairy farmers, but are equally relevant to dairy farmers throughout the world. Appendix 2 provides conversion factors to the standard metric system from a wide variety of systems used for describing weights and measures. Appendix 3 presents a conversion chart for the various Asian currencies used when discussing farm costs and returns associated with calf and heifer rearing. Appendix 4 presents typical expectation and evaluation forms developed for farmer workshops on improved young stock management.

Finally for ease of finding specific information, the Index lists all the key topics covered in the book and their relevant page numbers.

1.3 Role of the manual in farmer and adviser training programs

This manual is multipurpose in that it forms the basis of structured training programs in small holder dairying for a range of stakeholders in tropical dairy farming.

It is quite likely that many small holder (and even large-scale) farmers would find this book too technical and difficult to comprehend. In the first place, if English is not their mother tongue, unless they were well educated and/or travelled, very few would fully understand the level of English used. Secondly, the book is not simply a practical guide of how to undertake good young stock management. Rather, it is a technical manual about:

- why dairy calves and heifers suffer from poor stock management
- the symptoms of their suffering
- why their management needs to be modified
- how these practices can be improved for the betterment of the stock.

In other words, the book provides the theory behind the observations arising from poor management practices so farmers and advisers can understand the reasons why these practices need to be improved; that is, 'why unacceptable things happen' and 'how they should be changed' to ensure they do not happen again.

Most tropical countries have proactive programs to increase local supplies of milk, which require increasing numbers of well-trained workers to service the dairy industry. Consequently, educators from agricultural schools, universities and technical colleges need to be kept abreast of the latest technical developments and applications in dairy farming.

The key target audiences for this book are then:

- farmers who can understand this logical approach to improved young stock management
- farm advisers who assist farmers to improve their management practices
- educators, usually at technical level, who develop training programs for farmers
- educators, usually university level, who train dairy advisers in the basics of dairy production technology
- other stakeholders in tropical dairy production, such as agribusiness, policymakers and research scientists.

The text is written to be understood by advisers and tertiary students. As the trainers must ensure that other target audiences can comprehend their course material, they should select the most relevant sections to incorporate into basic programs for farmers. Each chapter has been written as a 'stand alone' document that can be individually downloaded from the internet, but there is some repetition, which has been kept to a minimum. For example there are several 'lists' of key farm practices for rearing milk-fed calves and weaned heifers, but each list has a specific purpose. Chapter 2 introduces all the key principles and processes for improved young stock management, while a checklist is provided in Chapter 16 for a structured assessment of current rearing practices on any farm. Chapter 17 briefly lists all the key elements of a training workshop for dairy farmers and advisers, while users of this manual are given the opportunity to answer 'yes' or 'no' to the 79 questions that make up the best management practices for calf and heifer rearing in Chapter 18.

To more fully understand the key factors influencing these basic steps of dairy production technology, they have been highlighted in diagrams at the beginning of each relevant chapter. This will provide extension workers and other dairy specialists with an 'easy-to-follow' checklist when using this manual to develop farmer training workshops.

For a book covering such a diversity of tropical dairy industries, every attempt has been made to present information from many countries. Much of the manual is based on my previous Australian calf and heifer books (Moran and McLean 2001, Moran 2002), but these key principles are relevant to any country with a dairy industry. Chapter 6 reviews published data of calf and heifer mortalities in the tropics, citing information from over 25 studies throughout the tropical world. Chapter 8 discusses restricted suckling and hand or machine milking, citing information from many Central and South American countries. Chapter 15 uses data from studies in Malaysia, Vietnam and Kenya to review the business of calf and heifer rearing. Chapter 17 is based on my experiences in developing farmer and adviser training programs throughout South and East Asia. Although the most common measures of currency seems to be the US dollar, Appendix 3 allows this to be converted into 12 other currencies to allow readers from many countries to more fully comprehend the costs and returns of their dairy farming. Throughout this book, milk intakes will be expressed either in volume (L) or weight (kg): because the density of whole milk is close to 1.0, the value is the same using both terms.

As previously mentioned, this is the fourth manual specifically written on tropical SHD farming. There is some inevitable overlap in topics covered within these books, so some repetition is inevitable because not all readers of this fourth manual will be familiar with all the technical aspects covered in the first three tropical dairy farming manuals.

1.4 The size of the heifer herd

It has been argued that, because every dairy farmer is different, there are just about as many systems of milk production as there are dairy farmers. For this reason, it is just not possible to categorise dairy farming into several discrete systems, with each system having a typical calf and heifer herd for which there would be an ideal set of herd management practices.

However, there can be some agreement on what constitutes small holder as against a large-scale dairy farming. In two of my previous books (Moran 2009, 2012), I used the following generally accepted descriptors to categorise dairy farms:

1. Small holder: up to 20 milking cows plus replacement heifers.
2. Semi-commercial: 20–50 milking cows plus replacement heifers.
3. Commercial: more than 50 milking cows plus replacement heifers.

Each of these farm types is usually located in a separate area with some degree of independence of supplies of forages and farm management. However, in certain countries, small holder farmers can be clustered together as 'colony farms', such as in centralised governed societies like China and even in dairy cooperatives in other countries such as Indonesia. With colony farming, small holders house their stock together in large dairy sheds, but are still responsible for feeding and maintaining their animals. These innovations require a large investment in buildings but they do allow small holders to own and manage their stock in a well-constructed durable shed with the benefits of magnitude of scale. This provides for communal activities such as forage production, machine milking and, of relevance to this book, heifer rearing. Either a single farmer takes charge of the rearing the replacement heifers for some or all of the small holder farmers in the 'colony', or the dairy cooperative may employ a specialist to rear them.

Assuming dairy farmers do not keep their male calves for breeding or dairy beef, they should aim to rear 25% annual replacements for their milking herd. Therefore each year, each small holder farmer with 20 milking cows should be rearing up to five heifers less than 12 months old (which are commonly called calves) together with another five weaned heifers over 12 months old (which are commonly called yearlings or heifers). Because the average herd size in many Asian countries is only two or three milking cows, the young stock herd may number only one or two.

Hence the calf and heifer herd size can be:

- one to two on many very small farms
- 10 on a 20 cow small holder farm
- 25 on a 50 cow semi-commercial farm
- 50 on a 100 cow commercial farm
- 100 or more on a colony farm.

In addition to young stock, non-lactating milking cows are also non-productive; that is, until they calve down again and recommence their lactation cycle. With non-productive dairy stock totalling 50% of the milking herd, young stock management cannot just be bulked together with general dairy farm management, but fully justifies a specific set of goals, procedures and a separate business plan.

1.5 Key performance indicators for rearing replacement heifers

1.5.1 Key performance indicators for components of heifer management

Poor heifer management is a major problem on many (if not most) Asian SHD farms (Moran 2005). Young stock receive insufficient attention because they do not generate income for many months. In addition, the first 3 months are the most expensive period in the life of any dairy cow and many farmers are just not prepared to invest in the calves' future. A low calf mortality rate indicates that early milk-rearing practices are adequate and allow for greater opportunity for economic and genetic improvement in the herd. When a heifer dies, there are fewer opportunities for culling unprofitable cows.

There are many hidden costs arising from poor management of the replacement dairy herd. The milking potential of small stunted animals that do not calve until 3 yr of age has been markedly reduced, while very high mortality and morbidity rates in calves during their milk feeding period represent an enormous waste of genetic potential in the dairy herd, as well as cash outlay (Moran 2002).

There are easily quantifiable benefits in having more newly calved heifers available to replace older unprofitable cows as both heifer and reproductive management improve. These benefits are:

- 1–2% more first-calf heifers for every month reduction in age at first calving
- 3–5% more first-calf heifers for every 10% reduction in calf mortality
- 2–3% more first-calf heifers for every month reduction in inter-calving interval.

Farmers should aim to rear 20–25% of their milking herd each year as replacements, to calve down for the first time by about 2 yr of age and produce at least five calves during their productive life. Realistic targets for tropical dairy systems (Moran and Tranter 2004) are:

- calf mortality to weaning: 4–6%
- heifer wastage rate from birth to second calving: 20–25%
- live weight at mating: 250–300 kg
- live weight at first calving: 400–500 kg (depending on breed type)
- age at first calving: 28–30 months.

Another good indication of heifer management is first lactation milk yield. This is expressed as a percentage of mature cow production, with a target of 80–85%. If first lactation yields are less than 75% of the mature equivalent, then the heifer-rearing program should be reviewed (Moran and McLean 2001).

On most dairy farms, heifers don't cover their rearing costs until they reach their second lactation. If they are culled or die earlier, they will leave the farm with an unpaid debt. A feed plan for heifers is as important as having a feed plan for milking cows. It is easy to let heifers drop down the priority list. Growing heavier, well-framed heifers is an investment in the future of the farm.

1.5.2 Key performance indicators for entire heifer herds

Replacement heifers are bred to allow for the culling of cows no longer suitable for the milking herd. Good heifer management is essential to provide sufficient animals for this to occur on a regular basis. The proportion of heifer calves that survive and grow well enough to become replacements depends on the many factors. These can be quantified as the proportion of:

- milking cows that actually conceive (the conception rate)
- those that produce a live calf (namely, do not abort during pregnancy or suffer neonatal death)
- those that are heifers (usually 50% of the viable calf drop, except when using sexed semen)
- those that survive until calving (namely, do not die during milk rearing and post-weaning)

Table 1.1. Measures of reproduction and calf rearing to produce replacements for a stable dairy herd (STOAS 1999)

Rearing system	A	B
Inter-calving interval (months)	12	18
Calving rate (%)	85	65
Stillborn calves (%)	2	5
Calf mortality from 0 to 24 months (%)	8	20
Non-pregnant heifers (%)	5	10
Heifer calves born (%)	36	15

- those that conceive as maiden heifers
- those that are suitable as milking cows in the herd (for example, are not culled because of poor temperament, poor udder conformation or because of lengthy illness).

STOAS (1999) compared reproduction and calf survival in two rearing systems (A and B) to calculate their relative replacement rates for a dairy herd with stable stock numbers (Table 1.1).

Assuming cows remain in the milking herd for up to four to five lactations, 20–25% should be replaced each year. From Table 1.1, the supply of 36% heifers from System A allows for the sale of young breeding stock or a higher culling rate to better address genetic improvements in the herd. Only one in every six or seven cows could be replaced annually in System B, which would hardly be enough to maintain herd numbers, let alone allow for much genetic selection. System A could then be considered as a set of key performance indicators.

With high ages at first calving (>30 months) and long inter-calving intervals (>15 months), it is very difficult to increase herd size through natural increases. That is why it is so important to seek the underlying causes of herds with high percentages of dry cows or a high proportion of heifers to cows. The most likely cause is poor feeding management, but there could be others, such as disease, heat stress or simply poor reproductive practices.

2

The two phases of young stock management

This chapter lists the key objectives during the two phases of heifer rearing.

The main points in this chapter

- The first phase is milk rearing, from birth to weaning.
- The second phase is post-weaning up until point of first calving.
- The key objectives of these two phases are listed.

There are two phases of young stock management on any dairy farm, namely:

- the milk feeding calf phase, from birth to weaning at 2–4 months of age
- the weaned heifer-rearing phase, from weaning to point of first calving.

The key objectives are presented in Figure 2.1 for milk-fed calves and Figure 2.2 for weaned heifers. This chapter simply lists the key objectives during the various steps in these two phases. The logic behind these objectives is discussed fully in the following chapters of this book. In my previous books for dairy farmers in Australia (Moran and McLean 2001, Moran 2002), I summarised the most important objectives as 'John Moran's golden rules for calf and heifer management' and these are presented in Appendix 1 of this book.

2.1 The milk feeding calf phase

The key principles of milk-rearing herd replacements are:

- ensuring healthy cows can give birth to healthy calves in a clean and comfortable environment
- providing suitable colostrum to allow adequate transfer of immunity
- supplying milk, fresh water and appropriate and timely supplements
- providing appropriate and clean housing
- minimising the risk of disease and disease spread

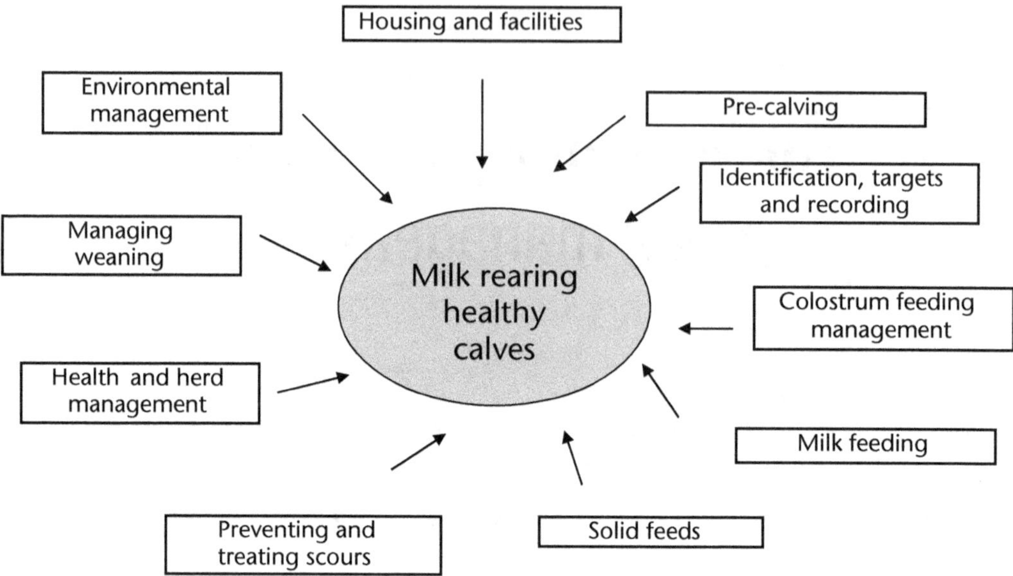

Figure 2.1. Key objectives during the milk-rearing phase of young stock management

- managing the weaning process
- instigating practices that reduce the risk of antibiotic and chemical residues
- ensuring calf and welfare requirements are met for any calves sold.

The process of milk rearing can be broken down to 10 steps, which together with some of the key objectives are described in Sections 2.1.1 to 2.1.10.

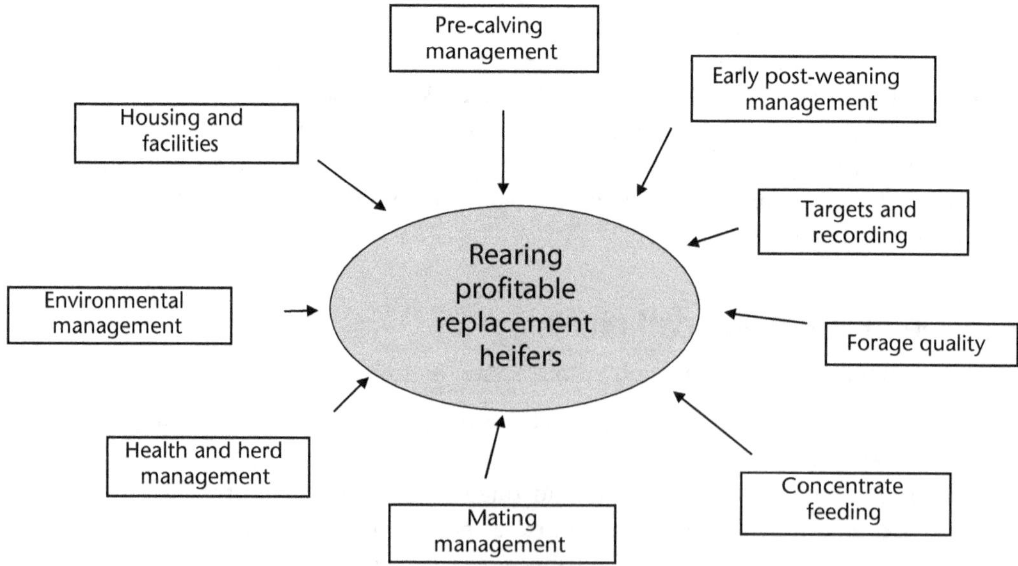

Figure 2.2. Key objectives in the weaned heifer phase of young stock management

The first phase is milk rearing the calf.

2.1.1 Pre-calving

- Select sires for ease of calving.
- Manage the transition period, before and after calving, to minimise metabolic diseases.
- Ensure that target mating weights are achieved before joining heifers.
- Implement a farm-specific vaccination program.
- Prepare facilities for calving, concentrating on space and hygiene.

2.1.2 Identification, targets and recording

- Record all details of the birth.
- Permanently identify each calf.
- Set targets for feeding, weaning and disease control.
- Decide on a protocol for culling very sick calves, following treatment.
- Establish a method for humane destruction, if required.
- Decide on whether to use individual or group pens.
- Develop a calf recording schedule (in a notebook or on a computer).
- Record all calf-rearing costs (including labour input) to calculate the total costs to weaning.

2.1.3 Colostrum feeding management

- Consider vaccinating the cows for local diseases to improve the colostrum quality.
- Preferably separate the cow and calf within a few hours and administer the colostrum by hand.
- If the cow and calf run together for 12 hr or more, ensure the calf has a good drink of colostrum from her dam.
- Ensure an adequate colostrum feeding program, including:
  - the quality of colostrum
  - the quantity of colostrum fed
  - quickly feeding the calf after birth, within the first 6 hr
  - the method of feeding colostrum
  - ensuring bull calves also receive adequate colostrum even if they are to be sold
  - continuing feeding colostrum, when available, for several days.

2.1.4 Milk feeding and access to drinking water

- Select the type of milk (whole milk or calf milk replacer – CMR).
- Be aware of the potential problems with very cheap CMRs.
- Decide on the method of feeding milk (buckets, teats, trough or restricted suckling)
- Seriously assess the benefits of an automatic calf feeding system for large calf-rearing enterprises.
- Do not dilute milk or colostrum with water.
- Decide on a feeding frequency (once versus twice each day).
- Decide on a warm or cold milk-feeding system.
- Be consistent each day (volume and temperature).
- Ensure good utensil hygiene when collecting and feeding milk.
- Provide clean drinking water from day 1 of life.
- Use hot water, detergents and sanitisers when washing feeding equipment.

2.1.5 Solid feeds

- Provide good quality concentrates (consider energy, protein and minerals).
- Understand the importance of adequate protein in the concentrates; milking cow concentrates are unsuitable due to low protein.
- Decide on a formulation (mixed on farm from base ingredients or commercially formulated).
- Provide to appetite from first week of age.
- Ensure fresh concentrate is on offer at all times.
- Do not mix concentrates with water.
- Monitor daily concentrate intakes to aid with the weaning program.
- Decide on forage supplements (preferably hay rather than fresh forage).
- Feed limited amounts of forages.

2.1.6 Preventing and treating scours

- Be aware that good colostrum feeding management is the key.
- Be consistent with all feeding and herd management.
- Minimise stresses on calves (overcrowding, climatic conditions, rough handling).
- Understand the difference between nutritional and infectious causes of scours.
- Understand types of scours and age when they can occur.
- Understand how to identify type of scours from symptoms (age, body temperature, faecal characteristics).
- Understand how to assess degree of dehydration in sick calves (sunken eyes, skin pinch test).
- Have good quality electrolyte fluid replacement on hand.
- Keep feeding milk but space out feeding times.
- Do not consider antibiotics until the type of infectious scours is identified by a veterinarian.
- Have a hospital pen for isolation of sick calves.

2.1.7 Health and herd management

- Dip the calves' navels in iodine solution soon after birth.
- Understand calf behaviour.
- Be confident with veterinarian support.
- Be aware of how to assist the veterinarian with follow-up treatments of sick calves.
- Purchase items for a 'calf nursing kit' (thermometer, stomach tube feeder and watch).
- Develop knowledge of local disease problems (such as pneumonia, joint-ill or bloat).
- Minimise faecal contamination of the calf-rearing area.
- Minimise exposure to infections.
- Identify, record and isolate all treated calves.
- Within group pens, don't mix calves of different ages.
- Plan for disease prevention rather than treatment.
- Consider routine vaccination against Clostridial diseases.
- If using antibiotics, record treatment dates and the withholding period to avoid sale of contaminated stock.
- If selling stock, ensure they are fit for travel and sale.
- Ensure excellent hygiene in calf pens and of milk feeding equipment.
- Be aware of important calf welfare issues.
- Record all instances and degree of health problems for later reference.
- Decide on protocol for other herd management (disbudding, vaccinations, and internal and external parasite prevention and treatment).
- Ensure newly introduced stock are kept in a separate quarantine area.
- Ensure all staff are aware of farm health protocols.
- Ensure any children wash their hands and change into clean clothes after leaving the calf-rearing area.

2.1.8 Managing weaning

- Because weaning is a process, not an event, it must be carefully planned.
- Decide on a weaning protocol (based on concentrate intake, age or weight or some combination of these).
- Preferably use concentrate intake and wean once calves are eating 750 g to 1 kg/calf/day.
- Immediately remove milk from the diet (don't dilute with water or gradually reduce intake).
- Note whether concentrate intakes quickly increase following weaning.
- Minimise early post-weaning stress, such as vaccinations and disbudding.
- Move weaned calves to other pens only after a few days.
- Continue feeding concentrates for many months even if forage quality is good.

2.1.9 Environmental management

- Protect stock from climatic stresses.
- Consider sprinklers and fans in very hot climates.
- Maintain a clean environment to minimise the disease risk.
- Ensure the rearing facilities do not cause undue stress on newborn calves.
- If cold weather is likely to be experienced, ensure the walls are solid to calf height.
- Provide artificial heating for sick calves in cold conditions.

2.1.10 Housing and facilities

- Consider individual calf cages for rearing calves for the first few weeks of life.
- If using individual pens, provide a slatted floor or comfortable bedding, such as straw or sawdust.
- Ensure each pen has containers for fresh water, milk, concentrates and, if being fed, forage.
- If using group pens, ensure no more than six calves per group, with sufficient floor space for each calf (at least 1.5 m^2/calf).
- Ensure good ventilation in the calf shed.
- Provide adequate lighting in the shed for night-time activities.
- Ensure hot water is readily available for cleaning purposes.
- Provide a small refrigerator for storing vaccines and other drugs in a secure area.
- Clean and disinfect each calf pen between batches of calves.
- Devise a good effluent disposal system for the regular cleaning of facilities.

2.2 The weaned heifer phase

The key principles of rearing weaned replacements heifers are:

- ensuring they grow well to achieve target live weights for mating and calving
- basing the ration on high-quality forages with concentrates specially formulated for growing heifers

The second phase is rearing the weaned heifer.

- supplying fresh water at all times
- providing appropriate and clean housing
- minimising the risk of disease and disease spread
- ensuring welfare requirements are met for any calves sold
- monitoring heifer performance to fine tune feeding management.

The aim of heifer rearing is to achieve maximum growth and development and earliest puberty at least cost. This ensures that maintenance costs are minimised, that there is the earliest possible return on the investment in the original animal and that the heifer can produce well during her first lactation. Any small growth check, due to sub-optimal feeding, is unlikely to have a permanent effect on her future productivity (unless it is severe), but can delay puberty: increasing rearing expenses hence decreasing profitability. The feeding of high-quality forages, including legumes, should ensure acceptable growth rates, but will invariably require additional high-energy and protein supplements and possibly additional forages to ensure good year-round live weight gains. Growing heifers may need to be housed to protect them from ectoparasites, such as ticks and biting flies and, in Central and southern America, from vampire bats during the night.

Tropical cattle do not grow as quickly in the tropics as do temperate dairy heifers in temperate regions. This is primarily because of the nutritional limitations of tropical forages, compounded by heat stress, which reduces appetite. Tropical dairy cattle also have lower mature sizes, hence less propensity for live weight gain.

In temperate areas, dairy heifers are first mated at 15–18 months, depending on mature size. However, in the tropics, the majority of heifers are too small, hence too immature, to breed at these ages. Live weight, rather than age, should be used as the criterion for mating heifers. Adequate live weights should be 200–225 kg for smaller and 290–315 kg for larger breeds. Following conception, heifers must continue to grow as well as produce a viable calf 9 months later. In addition, parasite control and routine vaccinations should continue during their rearing. Vaccinations for clostridia, anthrax, brucellosis, rinderpest, and foot and mouth disease should be routine, depending on the dairy region. During their last 2 months of pregnancy, additional feeding is required, usually with concentrates.

The process of rearing weaned heifers can be broken down to nine steps, which together with some of the key objectives for each step are presented in Sections 2.2.1 to 2.1.9. There is inevitably some duplication with the lists above for milk-fed calves.

2.2.1 Early post-weaning management

- Minimise stress immediately before and after weaning.
- Avoid moving calves out of milk-rearing pens for several days after weaning.
- Monitor individual concentrate intake after weaning to ensure rumen development is adequate.

2.2.2. Targets and recording

- Routinely weigh (or use chest girth tapes) to monitor changes in live weight during heifer rearing, say every 3 months.
- Use target live weights to modify feeding management, if required.
- Use body condition as an extra guide to heifer feeding management.
- Continue recording animal health treatments until first calving.
- Plan mating at target mating weights.
- Record mating times to aid with planning additional inseminations.
- Decide on protocol for culling very sick heifers, following treatment.
- Establish a method for humane destruction, if required.
- Develop a heifer recording schedule (in a notebook or on a computer).
- Record all heifer rearing costs (including labour input) to calculate total costs to first calving.

2.2.3 Forage quality

- Ensure the forage is of good quality.
- Provide adequate forage.
- Chop forages into small lengths.
- Consider wilting fresh grass to increase forage intakes.
- If possible, include legume forages to provide additional forage protein.

2.2.4 Feeding of concentrates

- Continue feeding concentrates until puberty and even after, depending on forage quality.

- Ensure concentrates contain adequate protein for heifer growth.
- Understand the importance of adequate protein in the concentrates; milking cow concentrates are unsuitable due to low protein levels.
- Decide on the formulation (from base ingredients or commercially formulated).
- Do not mix concentrates with water.
- Provide fresh drinking water at all times.
- Clean out the feed troughs at least once each day.

2.2.5 Mating management

- Ensure target weights are achieved before joining heifers.
- Seriously assess the benefits of synchronising oestrus in large groups of heifers.
- Consider mating heifers in specific seasons, particularly if summers are hot and humid.
- Ensure adequate infrastructure and support if depending on artificial insemination (AI) for mating heifers.
- Select bulls and semen on ease of calving.
- Consider access to bulls for natural mating following several cycles of AI.
- Ensure heifers are well fed and gaining weight during the mating period.

2.2.6 Health and herd management

- Develop a routine disease control protocol for heifer rearing. This should include vaccinating against diseases relevant to the location, control and prevention of external parasites as necessary.
- Develop skills in understanding stock behaviour.
- Be confident with veterinarian support.
- If using bulls for natural mating, ensure they have been assessed for fertility and treated to prevent spread of reproductive diseases.
- Plan for disease prevention rather than treatment.
- If selling stock, ensure they are fit for travel and sale.
- Be aware of important calf welfare issues.
- Record all instances and degree of health problems for later reference.
- Ensure newly introduced stock are kept in a separate quarantine area.
- Ensure all staff are aware of farm health protocols.
- Ensure any children wash their hands and change into clean clothes after leaving the calf-rearing area.
- Be aware of how to assist the veterinarian with follow-up treatment of sick calves.
- Purchase items for a 'calf nursing kit' (thermometer, stomach tube feeder and watch).
- Develop knowledge of local disease problems (pneumonia, joint-ill or bloat).

2.2.7 Environmental management

- Minimise heat (and cold) stress, as with milking cows.
- Protect stock from climatic stresses.
- Consider sprinklers and fans in very hot climates.

- Maintain a clean environment to minimise the disease risk.
- If cold weather is likely to be experienced, ensure the walls are solid to calf height.

2.2.8 Housing and facilities

- Ensure heifers have adequate space in their pens (at least 1.5 m^2/heifer).
- Ensure good ventilation in the heifer shed.
- Provide adequate lighting in shed for night-time activities.
- Provide a small refrigerator for storing vaccines and other drugs in a secure area.
- Clean and disinfect each pen between batches of heifers.
- Devise a good effluent disposal system for regular cleaning of facilities.

2.2.9 Pre-calving management

- Manage the transition period, before and after calving, to minimise metabolic diseases.
- Implement a farm-specific vaccination program.
- Prepare facilities for calving, concentrating on space and hygiene.

3

Digestion of feeds in the milk-fed calf

This chapter describes the various processes of digestion in the milk-fed calf.

The main points in this chapter

- The adult animal requires a fully functioning rumen to digest the fibrous feeds. The rumen is undeveloped in newborn calves, which depend on abomasal digestion until weaned off milk.
- Milk bypasses the rumen via the oesophageal groove where it forms a clot and is digested in the abomasum.
- Rumen development depends on the intake of solid feeds, which stimulate the rumen wall to absorb feed nutrients.
- Rumination, or 'chewing the cud', is a good sign of rumen development in milk-fed calves.
- The inclusion of roughage in the diet allows for earlier weaning. However, calves must consume high energy/protein concentrates for growth as well as rumen development.

If all calves could be reared by their natural mothers, there would be little need for this book. Most beef cows do a good job of rearing their own offspring, provided due care is paid to their feeding and health. The first essential of good husbandry in rearing calves is to keep them alive and fit enough to perform well later on (Moran 2002). To do this, farmers need to understand the development of the calf's digestive tract and the basic concepts of how calves digest their food. This is illustrated in Figure 3.1.

3.1 The calf digestive tract

An adult ruminant needs four functional stomachs to give it the ability to use the wide range of fibrous feeds available.

The reticulum and the rumen harbour millions of microbes, which ferment and digest plant material. The omasum allows for absorption of water from the gut contents. The abomasum or fourth stomach is the true stomach, comparable with that in humans and allows for acid digestion of feeds.

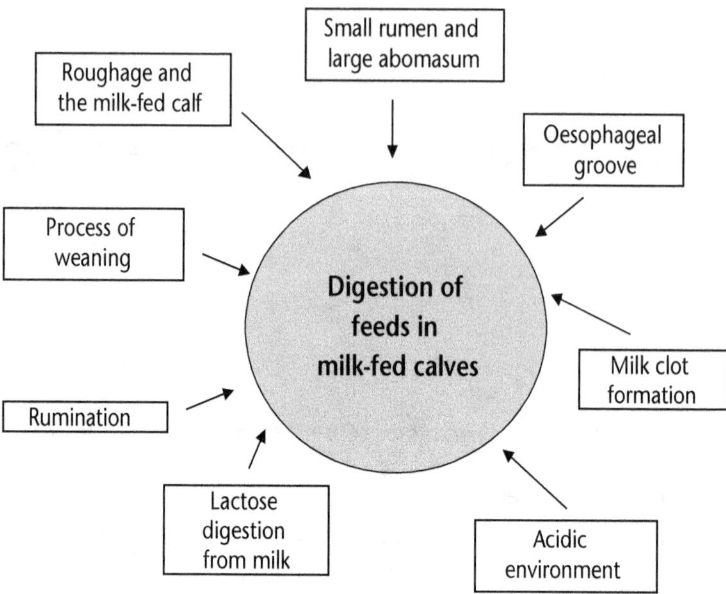

Figure 3.1. The key factors involved in digestion of feed in milk-fed calves

The very young calf has not developed the capacity to digest fibre and so the abomasum is the only functional stomach at birth. Both newborn and adult animals have a functioning small intestine, which allows for the alkaline digestion of feeds.

Figure 3.2 illustrates the anatomy of the stomachs and small intestine of a newborn calf. This schematic diagram shows the relative sizes of the four stomachs, the oesophageal groove, which runs from the oesophagus through the rumen to the abomasum, and the pyloric sphincter or valve at the bottom of the abomasum, which controls the rate of movement of gut contents into the duodenum.

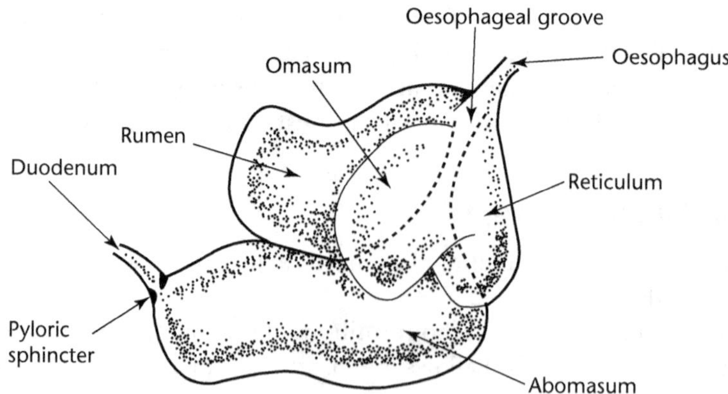

Figure 3.2. The four stomachs and duodenum of the newborn calf

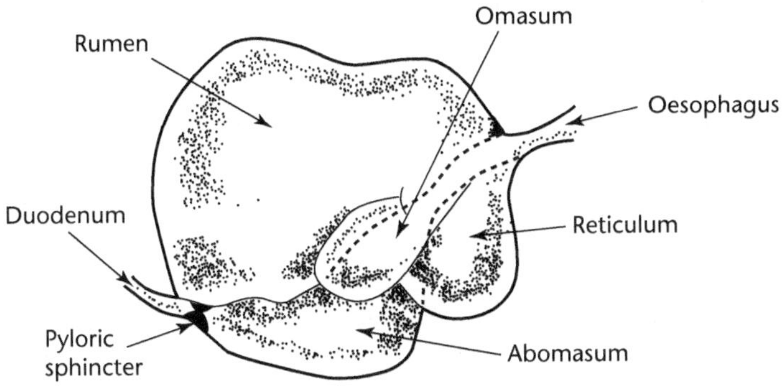

Figure 3.3. The four stomachs and duodenum of an adult cow

The omasum and abomasum account for about 70% of the total stomach capacity in the newborn calf. By contrast, in the adult cow, they only make up 30% of the total stomach capacity (Figure 3.3).

Digestion of feeds is aided by the secretion of certain chemicals called enzymes into the various parts of the gut. For example, calves produce the enzyme rennin in the abomasal wall to help digestion of milk proteins, while lactase is produced in the wall of the duodenum for digestion of milk sugar (lactose). These enzymes operate most effectively at different levels of acidity in the gut contents: acid in the abomasum and alkaline in the duodenum. To achieve this, the calf secretes electrolytes, or mineral salts, with the enzymes, to change the gut contents from one type to another.

The end products of digestion of the different components of feeds are absorbed through the gut wall into the bloodstream where they are taken to the different parts of the body for the animal's growth and development.

3.2 The milk-fed calf

Milk or calf milk replacer, whether sucked from a teat, or drunk from a bucket, is channelled from the oesophagus via the oesophageal groove into the abomasum. This groove is a small channel in the rumen wall that is controlled by muscles that allow liquids to bypass the rumen.

The groove is activated in response to different stimuli. It works well when calves suckle from their mothers or from teats, but sometimes does not work when they drink from a bucket. This appears to be a psychological condition in response to calves being separated from their mothers. Most calves can be trained by patient coaxing to drink quickly and well, responding to the new daily routine and the substitute mother in the shape of the calf rearer. When milk or milk replacer enters the abomasum, it forms a firm clot within a few minutes under the influence of the enzymes rennin and pepsin. This is the same process involved in making cheese or junket, using rennin to coagulate the milk protein. The clotting of milk slows down the rate at which it flows out of the abomasum, thus allowing for a steady release of feed nutrients throughout the gut and eventually into the bloodstream. It can take as long as 12–18 hr for the milk curd to be fully digested.

The enzymes acting on milk proteins require an acidic environment and this is provided by hydrochloric acid secretion into the abomasum. However, until the acid digestion is operating efficiently, which can take up to 7 days from birth, the only form of protein that can be digested is casein. There is no substitute for casein in the very young calf. Milk replacers containing other forms of protein cannot be digested until the calves are older.

Digestion of milk can be improved by including rennet, which can be obtained from cheese factories, or commercial calf milk additives for the first week or so. These can provide additional acids to reduce abomasal pH, and enzymes and specific bacteria to increase the rate of breakdown of the milk curd. Such additives are called probiotics, in that they aid in normal digestive processes. Research has not always found these to improve calf performance and health greatly, although they are more likely to be beneficial when calves are suffering from ill health.

Any milk from a previous feed is enveloped in this newly formed clot. Liquid whey protein and lactose are rapidly separated from the milk curd and pass into the abomasum. The milk fat embedded in the milk curd is broken down by another enzyme, lipase. This is secreted in the mouth in saliva and incorporated when milk is swallowed. Teat feeding, rather than bucket feeding, seems to produce more saliva and hence more lipase. Further digestion of the milk protein and fat occurs in the duodenum with the aid of enzymes produced in the pancreas.

Lactose, which is quickly released from the milk curd in the abomasum, is broken down to glucose and galactose and these are absorbed into the bloodstream to form the major energy sources for young calves.

Fats are broken down into fatty acids and glycerol for absorption and use as energy, while proteins are broken down into amino acids and peptides for absorption and use as sources of body protein.

Starch, such as from cereal grains, is an important source of energy in older calves, but calves in their first few weeks of life cannot digest starch.

The abomasum is not acidic until the calf is 1–2 days old and this has advantages and disadvantages. The major advantage is that the immune proteins in colostrum cannot be digested in the abomasum, so are absorbed into the bloodstream in the same form as when produced by the cow. This ensures their role as antibodies to protect against infection. The low acidity of the abomasal contents in the newborn calf constitutes a potential risk from the bacteria (and probably viruses) taken through the mouth. These will not be killed by acid digestion and they can pass into the intestines where they can do the most harm. All calves pick up bacteria in the first few days of life and this is essential for normal rumen development. However, some of the first bacteria to colonise the gut can also cause scouring. Provided the calf has drunk colostrum, the maternal antibodies can control the spread of these more harmful bacteria.

The milk-fed calf must then produce an acid digestion in the abomasum and an alkaline digestion in the duodenum. This is achieved by the production of electrolytes in the gut wall.

Calves suffering from scours due to nutritional disturbances or bacterial infections can lose large amounts of water and electrolytes in their faeces. These must be replenished as part of the treatment for scours.

Colostrum is the first milk produced by newly calved cows. Not only does it provide essential feed nutrients, it supplies maternal antibodies that allow passive transfer of immunity against diseases of calves. Recommendations for colostrum feeding are discussed in Chapter 5.

3.3 Rumen development and the process of weaning

After calves are weaned, the cost of rearing declines markedly. Feed costs are lower, labour inputs are reduced and incidence of ill health is less. It makes economic sense to wean calves as soon as is reasonable. However, the calf is forced to undergo several dramatic changes, namely:

- The primary source of nutrients changes from liquid to solid.
- The amount of dry feed the calf receives is increased.
- The calf must adapt from a monogastric to a ruminant type of digestion, which includes fermentation of fibrous feeds.
- Changes in housing and management often occur around weaning, which can add to stress.

At birth, the rumen is a small and sterile part of the gut, but by weaning it must become the most important compartment of the four stomachs. It must increase in size, internal metabolic activity and external blood flow. The five requirements for ruminal development are:

- the establishment of bacteria
- plenty of liquids
- the outflow of material (muscular action)
- the absorptive ability of the tissue
- a substrate to allow bacterial growth, such as recycled minerals, as well as feed nutrients.

Prior to consumption of solid feeds, bacteria exist by fermenting ingested hair, bedding and milk that flows from the abomasum to the rumen. Most water entering the rumen comes from free water (actual water, not water contained in milk or milk replacer solution). Milk will bypass the rumen via the oesophageal groove, whereas free water will not.

The rumen develops from a very small organ in newborn calves (1–2 L) into the most important part of the gut (25–30 L) by 3 months of age. It can enlarge very quickly during the first few weeks of life, given the right feeding management. Rumen growth only occurs under the influence of the end products of rumen digestion, which result from the fermentation of solid feeds by the rumen microbes. Development largely occurs through growth of rumen papillae on the rumen wall (leaf-like structures on the internal surface), which increase the surface area of the rumen and hence its ability to absorb these end products of digestion.

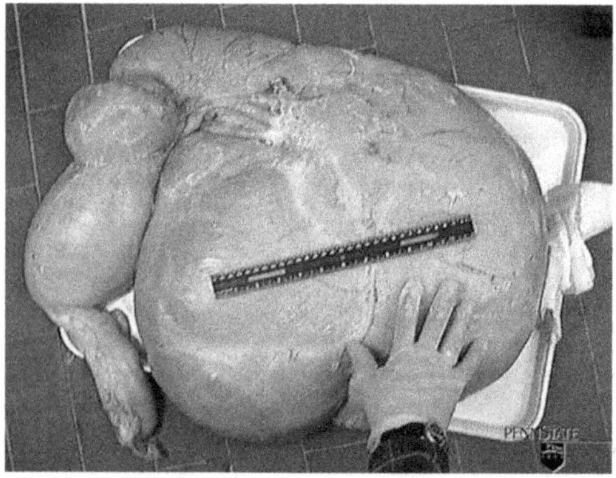

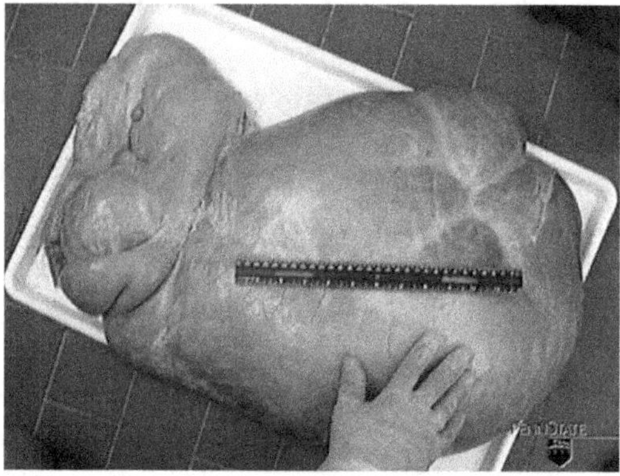

The dissected rumen of a 12-week-old calf fed entirely on milk and hay (left) or on a combination of milk, hay and grain (right). Note the difference in rumen capacity as a result of grain feeding. (Photos used with permission.)

The dramatic effects of feeding cereal grain, in addition to milk and hay, are visually depicted in changes in rumen size and rumen wall development of a 12-week-old calf in the photos below.

The rumen's capacity and the intake of solid feed are closely related. Rumen development is very slow in calves fed large quantities of milk. The milk satisfies their appetites so they will not be sufficiently hungry to eat any solid feed.

Rumination, sometimes called 'chewing the cud', can occur at about 2 weeks of age and is a good indication that the rumen is developing. Solid feeds and rumination both stimulate saliva production and this supplies chemicals such as urea and sodium bicarbonate to produce the substrates for bacterial growth.

When weaning calves early, it is important to limit both the quantity of milk offered and its availability throughout the day. It is also essential to provide solid feeds. Roughages (of low or high quality) should be offered in combination with high-quality concentrates. Roughages stimulate rumen development, while concentrates supply feed

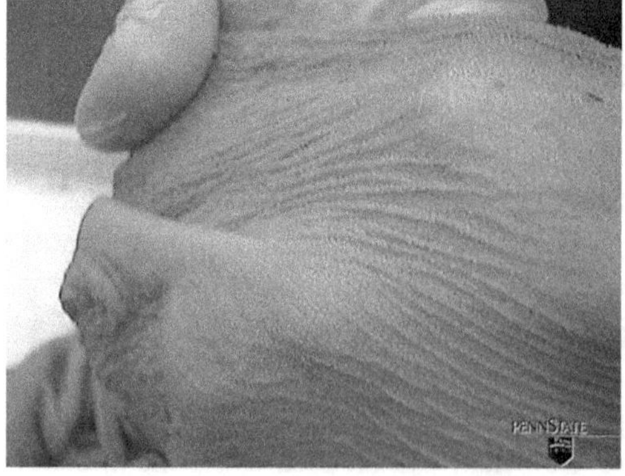

The rumen wall of a 12-week-old calf fed entirely on milk and hay (left) or on a combination of milk, hay and grain (right). Note the difference in papillae development as a result of feeding grain. (Photos used with permission.)

nutrients not provided by the limited quantities of milk offered. Without the concentrates, calf growth is slow but the rumen still develops, resulting in undesirable pot-bellied animals.

Urea supplies nitrogen for the microbes, while sodium bicarbonate acts as a rumen buffer, helping to maintain a steady pH in the rumen contents. This is particularly important when calves eat large quantities of cereal grains later in life because the rumen microbes can produce a lot of lactic acid during fermentation.

Grain poisoning or acidosis occurs when lactic acid levels are excessively high and become toxic to the rumen microbes and eventually to the animal. As well as the end products that are absorbed through the rumen wall, microbial fermentation produces the gases carbon dioxide and methane, and these are normally exhaled. When something prevents the escape of these gases from the rumen, bloat can result at any stage of life.

3.4 The role of roughage in the weaning process

The inclusion of roughage in the diet improves intakes, performance and allows for earlier weaning.

Fresh pasture is not the ideal roughage source for milk-fed calves because it has too little fibre and a low feed energy density. Its high water content limits its ability to provide adequate feed energy for the rapidly growing animals. Until their rumen capacity is larger, young calves just cannot eat enough pasture for good growth, unless it is very high in quality.

Calves reared on restricted milk plus concentrates display good rumen function by 3 weeks of age and have sufficient rumen capacity for weaning by 4–6 weeks of age. However, if the diet was restricted milk plus high-quality pasture, rumen capacity may not be sufficient for weaning until 8–10 weeks of age. Even then, growth rates would be lower in calves weaned onto pasture alone because of insufficient energy intake due to the physical limitations of rumen capacity.

If too high in quality and fed *ad lib* (that is fed to appetite), calves seem to prefer the roughage to concentrates, leading to a reduced intake of feed nutrients and slower growth. When clean cereal straw and concentrates are both fed *ad lib* together with limited milk, calves will eat about 10% straw and 90% concentrates. Without the roughage and the resulting rumination, rumen development will be slower due to insufficient saliva and end products of fibre digestion.

This chapter highlights the importance of encouraging early rumen development through feeding management to reduce total feed costs and allow the calf to move from the milk-fed calf phase to the weaned heifer phase, when it becomes less dependent on the rearer for feed, disease and general herd management.

4

The nutrient requirements of calves

This chapter describes the various feed nutrient calves require for normal growth (Figure 4.1).

The main points in this chapter

- Water is essential for all living animals. Weaned calves require 10–15 L/day with up to 25 L/day on hot days.
- Energy is required to maintain body temperature and to support normal body function. Energy is best described as metabolisable energy (ME), with requirements quantified in MJ of ME/day.
- Milk is digested in the abomasum with much higher energetic efficiency than are solid feeds in the rumen.
- Proteins are required to maintain normal body processes, repairing tissues and forming blood and also laying down muscle. Protein is quantified in g/day or percentage of feed dry matter.
- Rumen development in the milk-fed calf depends on its intake of solid feeds that contain fibre.
- Calves also require small amounts of minerals and vitamins. The most important minerals are calcium, phosphorus and magnesium.

The three most essential nutrients for calf growth and development are water, energy and protein. Fibre, minerals and vitamins are also needed, but play a smaller role.

4.1 Water

Water is essential for all living animals and it is good husbandry to provide calves with as much fresh, clean water as they want. Weaned calves can drink 10–15 L/day and up to 25 L/day on hot days. For optimum feed efficiency, milk-fed calves need to drink 4 L of water for every kg calf concentrate they eat.

Milk contains 87–88% water, which should be sufficient for normal body requirements. Milk-fed calves will not suffer from the absence of extra water unless they

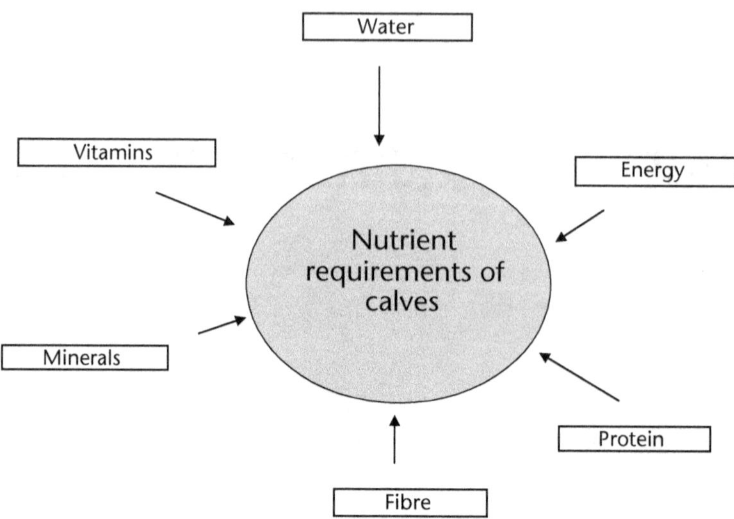

Figure 4.1. The nutrient requirements of calves

are exposed to heat stress. However, as soon as they start eating solid feeds, particularly dry feeds like hay or straw, calves require continuous or regular access to fresh water. This simple practice will increase their intake of solid feeds and so reduce their age at weaning.

To standardise the description of feed intake, it is usually expressed in terms of dry matter (DM). This is easily measured by placing feeds in an oven at 100°C for up to 48 hr. Animals eat quite similar amounts of DM no matter what type of feed is offered.

Continual access to fresh drinking is essential for all dairy stock.

Maximum DM intake is directly related to live weight. Setting a maximum value of 2.5–3.0% live weight per day, a 100 kg weaned calf will eat about 3 kg DM/day while a 200 kg calf would eat up to 6 kg DM/day.

4.2 Energy

Energy is needed to maintain body temperature and to support normal body functions. This is known as the maintenance energy requirement. Any energy consumed that is surplus to this basic need is available for growth or the laying down of muscle and fat, which leads to live weight gain. Maintenance energy requirements increase with live weight.

Energy requirements and the available energy in feeds are measured in units called joules, and more commonly in kilojoules (1000 joules) or megajoules or MJ (1 million joules).

Only a portion of the total (gross) energy in feeds becomes available to the calf following its digestion. Undigested energy is lost in the faeces, while a small portion of the digested energy is lost through rumen fermentation and also in the urine.

The remaining useful productive energy is called the metabolisable energy or ME. This is the conventional measure of the energy requirements of the calf (in MJ of ME/day) and also the energy content of different feeds (in MJ of ME/kg dry matter or MJ/kg DM).

Because milk is a high-quality feed that is digested efficiently in the abomasum, its energy value to the calf is considerably greater than that of solid feeds digested in the rumen. More than 90% of the gross energy in milk ends up as ME compared with only 50–60% of the gross energy in hay and concentrates.

The energy requirements for growth increase with age and live weight. They also vary with the energy content of the feed. High energy feeds, such as milk and concentrates, are used more efficiently for growth than low energy feeds, such as medium quality pasture or hay. Because the maintenance ME requirement is relatively constant for a particular live weight, the faster an animal grows, the higher the proportion of the total ME intake is available for growth, and therefore the more efficiently the feed is used by the calf.

For example, an 80 kg weaned calf growing at 0.5 kg/day requires 44 MJ for each kg of weight gain, and will only use 32% of its ME intake for growth. If growing at 1.0 kg/day, it requires only 31 MJ for each kg of weight gain and will use 52% of its ME intake for growth. Furthermore, it will only take half the time to achieve the same target live weight gain. The slower growing calf will then require considerably more ME than is apparent from the difference between these two ME requirements for growth. This highlights the greater value of any high-quality feed to achieve improved growth rates.

Calculations on the energy requirements of milk-fed calves differ from those for calves with developed rumens. A milk-fed calf weighing 100 kg and growing at 0.5 kg/day requires a total of 21 MJ of ME per day, 4 MJ/day less than if that calf was weaned. For calves fed milk, about 20% of the gross energy in milk is retained in the body of a 2-week-old calf growing at 0.3 kg/day, and this can increase to 26% for 14-week-old veal calves fed milk replacer and growing at 1.2 kg/day. However, only 11% of the gross energy intake is retained in the bodies of 6-month-old calves already weaned onto a diet of hay and concentrate, and growing at 0.6 kg/day.

In other words, solid feeds digested in the rumen are only used about half as efficiently by the growing calf as milk digested in the abomasum. However, the energy in milk can cost up to four times that in concentrates, making early weaning onto solid feeds considerably cheaper than milk feeding. This is discussed further in Chapter 15.

The reason that feed energy is the most important nutrient in any diet is because the efficiency of feed energy conversion into animal product is so low. Less than a quarter of the energy in milk fed to calves is retained by the animal and this falls to only one tenth of the solid feed when the calf is weaned. Once energy is used by the animal, some of it is lost as body heat, while some of the other feed nutrients are recycled within the animal.

Practical ration formulation is based on the principal of selecting the most appropriate ration ingredients to meet the animal's energy needs at the lowest cost. Requirements for the other nutrients, such as protein, fibre and minerals, can then be met by adjusting the concentration of the ration ingredients so that none of these limit animal performance. Table 4.1 shows examples of calculations on the energy and protein requirements for weaned calves to achieve various growth rates at different live weights. This table was first presented by Webster (1984) using standard feeding tables published in the UK by Agricultural Research Council (1980). In Asia, the energy and protein requirements for ruminants are often calculated using the US National Research Council (1989) standards.

Table 4.1 shows that the ME intakes of calves achieving 1.0 kg/day weight gain are only about a third higher than for calves growing at 0.5 kg/day, but about double those for maintenance. This table also presents the required ME and crude protein (CP) contents of any ration to achieve growth rates of 0.5 or 1.0 kg/day.

4.3 Protein

Proteins are required by the calf to maintain biological processes on a daily basis, as well as repairing tissues and forming blood. Proteins are also an integral part of growth, such as the laying down of muscle. Most protein synthesis takes place in other body tissues such as the liver and gut wall, which are actively concerned with processing nutrients to meet the requirements of the body. These metabolic functions include such things as synthesis of enzymes and hormones, cell division and cell repair, and so require a continuous supply of different types of protein and energy.

The building blocks of all proteins are amino acids. There are more than 20 specific amino acids needed by livestock. Feed protein is broken down by digestion into its individual amino acids, which the calf absorbs and then resynthesises for its maintenance and growth.

The precise needs for specific amino acids are well known in non-ruminants such as pigs and poultry but there is little information available on the amino acid requirements of calves and adult ruminants. Because the milk-fed calf depends entirely on its diet to supply them, the amino acid composition of whole milk would probably match its specific requirements.

Table 4.1. Requirements of weaned calves for metabolisable energy (ME), rumen degradable protein (RDP) and undegradable dietary protein (UDP) at different live weights and for different growth rates (prepared by Webster 1984)

	Live weight (kg)		
	80	140	200
Maximum DM intake (kg/day)	2.4	3.6	4.8
ME requirements (MJ/day)			
Maintenance (M)	15	23	30
M + 0.25 kg/day gain	18	27	36
M + 0.5 kg/day gain	22	32	42
M + 0.75 kg/day gain	26	38	48
M + 1.0 kg/day gain	31	43	55
Minimum dietary ME content (MJ/kg DM)			
0.5 kg/day gain	9.2	8.9	8.7
1.0 kg/day gain	12.9	11.9	11.5
Crude protein requirement (g/day)			
0.5 kg/day gain RDP	170	250	330
UDP	130	120	110
1.0 kg/day gain RDP	240	335	430
UDP	200	180	150
Minimum dietary crude protein content (%)			
0.5 kg/day gain	12.5	10.3	9.2
1.0 kg/day gain	18.3	14.3	12.1

Rumen microbes synthesise many of these amino acids in the older calf, during the formation of microbial protein, while others are provided by the undegraded protein in the diet. Although the essential role of amino acids is in forming proteins, they can be used as sources of energy when they are in excess to the calf's protein requirements. In other words, if more protein is fed than required, it is used for energy in much the same way as the starch in cereal grains or other energy sources. However, because protein is usually more expensive than energy to supply in feeds, it is important to provide only that amount required when formulating rations.

The element nitrogen (N) is an essential constituent of all proteins, present at about 16% of the DM, though varying slightly with different proteins. That is why when feeds are analysed for protein, the total N content is measured then multiplied by 6.25 (or 100/16) to give the level of crude protein, or 6.38 where milk products are concerned. This is a good measure of the capacity to provide amino acids in many feeds, but, in others, much of the crude protein is in the form of non-protein N, usually simple compounds such as urea. All feeds contain some proportion of their N as non-protein N.

Adult ruminants can benefit from this non-protein N because, when it is broken down by the rumen microbes into ammonia, they re-use it to synthesise their own microbial protein. In calves up to 6 months old, the crude protein should be mainly in the form of true protein. The nearer the composition of a feed protein is to that of the protein in calf weight gains, the more efficiently it can be used by the calf for growth because the supply of amino acids will more closely match its requirement. In other words, there will be less likelihood of any amino acids limiting calf performance or of excess amino acids being wasted as protein sources. The extent to which the true protein is broken down by microbial action depends on its vulnerability to microbial attack and the length of time it spends in the rumen.

Animal proteins, such as fish meal, are more valuable to calves than plant proteins because their amino acid make up more closely matches those of the rapidly growing calf. This is called the biological value of the protein for animals. However, it is now illegal to use animal proteins to feed ruminants in most countries because of the concern that they are associated with the spread of certain diseases such as 'mad cow disease'.

Dietary crude protein is further described in terms of two constituents, namely rumen degradable protein (RDP) and undegradable dietary protein (UDP). The RDP, which includes all the non-protein N and some true protein in the diet, is broken down in the rumen and then resynthesised into microbial protein at a rate determined by the energy metabolism of the rumen microbes. Two forms of dietary protein escape into the abomasum: namely UDP and microbial protein.

There is a fairly constant relationship between the amount of microbial protein produced and the availability of ME. If there is more RDP than available energy, the excess N that is already converted into ammonia is not recaptured by the microbes. This is absorbed through the rumen wall and converted into urea in the liver. Much of this blood urea is excreted in the urine and therefore wasted, although some is recycled back into the rumen as salivary urea.

The best way to ensure that feed protein is efficiently used by calves is to supply it in the form of UDP; that is, feed protein escaping rumen fermentation that passes directly to the abomasum for acid digestion.

The requirements for amino acids depend on the rate at which the calf is producing new tissue – its growth rate. As growth rate increases so to does its requirements for RDP and UDP (in g/day) and its total dietary protein content (as % DM). These are shown in Table 4.1 for calves growing at 0.5 and 1.0 kg/day.

These protein requirements were calculated for a crossbred Friesian steer calf. Bull calves and calves from larger European beef breeds require an additional 10–15% UDP. The table also converts these RDP and UDP values to more practical units, namely the minimum crude protein of the diet and the optimal degradability of the protein (calculated as the percentage of RDP in the total crude protein). Younger, lighter calves require higher dietary crude protein levels and need more of their protein as UDP (they have lower optimum protein degradabilities).

There is also evidence suggesting that the degradability of dietary protein can influence the composition of live weight gain. For example, if calves consume rations providing adequate energy and total crude protein, but UDP intakes are below recommended levels, growth rate may not be reduced but more of the live weight gain would be as fatty tissue and less as muscle. This has important implications for growing dairy heifer replacements because excess fat in the developing udder can reduce the potential for that udder to produce milk in later life. Early muscle growth is important for dairy beef calves to achieve target live weights for slaughter at set criteria of carcass fatness.

The supply of UDP, as opposed to RDP, in the diets of milk-fed calves is not important because liquid feeds already bypass the rumen digestion, through the oesophageal groove (see Chapter 3), and so supply the necessary UDP for growth and development. In fact, rumen development requires rumen digestion and so a supply of RDP in solid feeds will be beneficial to milk-fed calves. Because all feeds must enter the rumen in weaned calves, the type of feed protein is important. Most vegetable proteins are highly degraded, while animal proteins are more protected against rumen degradation. Extra processing – for example heating or the addition of chemicals such as formaldehyde – can reduce the degradability of feed proteins.

Increasing the supply of energy to the weaned calf will increase the amount of RDP that the rumen microbes can resynthesise into microbial protein. In other words, the balance of RDP and UDP is not constant for any type of feed but will vary depending on the other ingredients in the total ration.

Feeds high in UDP are processed protein meals such as cottonseed or canola meal. Feeds with high RDP/ME balances are fresh and conserved pastures, protein meals and urea. Feeds with low RDP/ME balances include cereal grains, maize silage and cereal straws. When formulating a ration for weaned calves, it is important to balance the feeds for both protein and energy.

4.4 Fibre

Highly fibrous feeds stimulate saliva production during chewing and rumination. The saliva provides urea and minerals, such as sodium bicarbonate, that help maintain normal rumen microbial growth and development.

Fine grinding of feeds changes the physical nature of the fibre, but not its chemical analyses, and this can reduce its effect on rumen development. Mixing roughage with concentrates to assure consumption of both feeds without separation may require the reduction of particle size to the point that the physical abrasion (or 'scratch factor') in the roughage has lost its beneficial effects. The initial introduction of solid feeds should contain from 10 to 20% of the DM as roughage, with the particle size maintained as large as possible.

Dietary fibre can be quantified in various ways. Traditionally it was described in terms of crude fibre (CF), but this is being replaced by two terms: namely acid detergent fibre (ADF) and neutral detergent fibre (NDF). ADF is made up of the cellulose and lignin in the feed, while NDF also includes an additional fibre source, namely hemicellulose. NDF then quantifies the cell walls in the forage and is discussed in Chapter 10.

4.5 Minerals and vitamins

The two minerals of most importance to growing calves are **calcium** (Ca) and **phosphorus** (P) because both are required for bone development. These minerals also have other, more dynamic functions such as in muscle function (Ca) and energy metabolism (P).

One of the earliest signs of deficiencies of these major minerals is poor growth and poor appetite. As with most mineral and vitamin deficiencies, these early signs are not

Table 4.2. Requirements of weaned calves for major minerals at different live weights and for different growth rates (prepared by Webster 1984)

Mineral	Growth rate (kg/day)	Live weight (kg)			
		100		200	
		g/day	% in DM	g/day	% in DM
Calcium	0.5	12	0.42	14	0.30
	1.0	21	0.75	24	0.50
Phosphorus	0.5	6	0.20	8	0.20
	1.0	11	0.40	13	0.30
Magnesium	0.5	3	0.10	4.8	0.10
	1.0	4.2	0.20	6	0.15

very specific; the calves simply do not appear to be doing well. This can also apply to calves suffering from parasitism or infectious diseases but, once these are eliminated, calves must be provided with the deficient nutrient before they respond. If they do not respond, then something else is wrong.

Calves do not possess what is often called 'nutritional wisdom'; they have no innate ability to select feeds to satisfy any particular nutrient craving. The only possible exception is sodium – cattle can sense the presence of sodium in rock salt or in drinking water at incredibly low concentrations. Despite what producers may be told, the intake by cattle or calves of mineral blocks bears little relationship to their mineral requirements. Deficiencies of calcium and phosphorus in milk-fed calves are rare. However, they can occur after weaning if calves are fed unbalanced diets.

The other major minerals for calves are **magnesium, sodium** and **potassium**. Deficiencies are unlikely except as complications of diseases that lead to scouring. They should be included in electrolyte solutions given as part of the treatment for scours. Table 4.2 shows the requirements of growing calves for major minerals, as summarised by Webster (1984), both as requirements in g/day and as minimum dietary DM concentrations.

Higher levels of dietary minerals are required to achieve growth rates of 1.0 kg/day compared with 0.5 kg/day. Most standard feeding tables have mineral concentrations of the available feeds and from these it is possible to determine whether additional mineral premixes should be included in calf diets.

Calf rearers can generally assume that purchased milk replacers and concentrate mixes contain the correct level of minerals for normal calf development. It is rare for problems to arise through mineral deficiency or imbalance, but as higher animal performance is sought by calf rearers, certain minerals may become limiting. One example of such an imbalance could arise from the low calcium content of most cereal grains when fed in large amounts to maximise growth rates.

If calves are grown for veal, it is important to monitor the level of **iron** in the diet. Calves are born with low reserves of iron and the amount of iron is very low in whole milk. Additional iron supplements can increase both blood haemoglobin levels and growth rates in young calves.

The colour of meat in calves is largely influenced by its iron status. White veal is very pale because the calf becomes anaemic and lacks those meat and blood pigments that are high in iron. Pink veal allows for higher intakes of iron through concentrate feeding, but they should be kept low enough to ensure pink rather than red coloured meat. White veal is normally produced on a diet entirely of whole milk or milk replacer, although veal producers are now being pressured to include solid feeds in white veal diets.

Calves are born with very low reserves of **vitamins A, D and E** and hence are very dependent on colostrum to supply these vitamins. Most milk replacers have enhanced levels of vitamins because of their importance to calf health. The milk-fed calf is also unable to synthesise its requirements for the complex of B vitamins and these are normally added to milk replacers. Once the calf has a functioning rumen, it is capable of supplying its own B vitamins, and these are not normally added to concentrate mixes.

5

The importance of colostrum to newborn calves

This chapter discusses the management of colostrum feeding in the first few hours of a calf's life.

The main points in this chapter

- Calves are born with no immunity against disease, so must depend on their dam to provide passive immunity through colostrum: the first milk produced after birth.
- The essence of good colostrum feeding management is the 3 Qs; quality, quantity and quickly.
- Quality is essential to ensure there are sufficient antibodies (against diseases) in the colostrum for absorption by the calf.
- Quantity to ensure the calf ingests sufficient of these antibodies.
- Quickly means the calf's digestive tract is still receptive to absorbing them.
- Modern day dairy cows do not produce good-quality colostrum and many calves will not drink soon after birth. Therefore it is better to stomach tube each newborn calf to ensure it can acquire passive immunity against diseases.
- The level of circulating antibodies in the calf's blood has a direct influence on its disease resistance, its performance as a mature cow and the financial contribution to the farm's profit.

Calves need to receive individual attention and care immediately following birth. There is a direct link between good calf care and improved milk production and longevity in the milking herd. Care of the newborn calf is discussed in Chapter 14 of this manual.

Calves are born with no immunity against disease. Until they can develop their own natural ability to resist disease, through exposure to the disease organisms in their surroundings, they depend entirely on the passive immunity acquired by drinking colostrum from their dam (see Figure 5.1).

Colostrum is the thick, creamy-yellow, sticky milk first produced by cows initially following calving, and contains the antibodies necessary to transfer immunity onto their

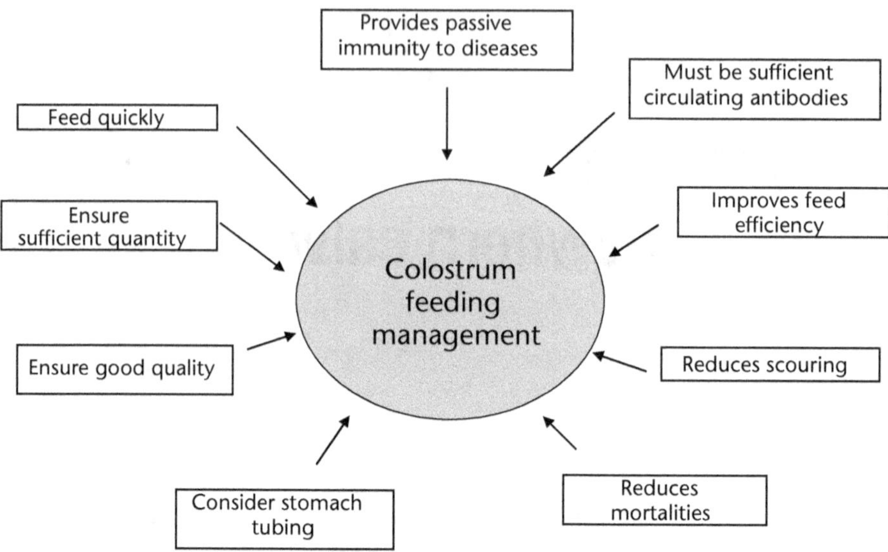

Figure 5.1. The key factors and benefits of colostrum feeding management in newborn calves

calves. It is essentially milk reinforced with blood proteins and vitamins. It has more than twice the level of total solids than in whole milk, through boosted levels of protein and electrolytes. It also contains a chemical allowing newborn calves to use their own fat reserves to immediately provide additional energy.

Newborn calves have no immunity against disease.

The calf should be assisted for its first suckle from its dam.

The concentrations of protein and vitamins A, D and E in colostrum are initially about five times those of whole milk, with a protein content of 17–18% compared with milk's 2.5–3.5%. However, within 2 days these are little different from those in whole milk. The levels of vitamins in colostrum are dependent on the vitamin status of the cow. The blood proteins transfer passive immunity from mother to offspring through maternal antibodies or immunoglobulins (Ig).

The chances of calves surviving the first few weeks of life are greatly reduced if they do not ingest and absorb these antibodies into their bloodstream. It takes far fewer disease organisms to cause disease outbreaks in such calves than if they can acquire immunity from their dam. Calves without adequate passive immunity are four times more likely to die and twice as likely to suffer disease, than those with it. Furthermore in certain situations, blood levels of antibodies in heifer calves are directly related to their milk production in later life.

The term colostrum is generally used to describe all the milk produced by cows up to 5 days after calving, until it is acceptable for use by milk factories. However, a more correct term for milk produced after the second milking post-calving is transition milk (Moran 2002). This milk no longer contains enough Ig to provide maximum immunity to calves, but still contains other components, which reduce its suitability for milk processing. Milk factories can now test for and penalise farmers who include transition milk in their milk vat. Because it has no market value, transition milk should all be fed to calves to reduce their total feed costs. However, it must be stressed that the immune properties of this pooled milk are much reduced once first milking colostrum is diluted with that from second or later milkings.

When considering colostrum feeding to dairy calves, it should be appreciated that modern milking cows are vastly different to the primitive, feral cows from which they evolved thousands of years ago. Their udders are much larger and often hang too low for easy suckling by their offspring. They produce vastly greater quantities of milk, which means that their first and second milking colostrum is much more dilute than is desirable for optimum quality. Furthermore, because mothering ability has little relevance on dairy farms and has probably been bred out of cows, they may be less likely to want to suckle their progeny immediately after birth. This is still not the case with beef cows, where unassisted suckling is a highly efficient means of passive transfer of immunity in beef calves. These natural methods are less effective in dairy herds, meaning that farmers often have to rely on so-called less natural techniques.

5.1 Current recommendations on colostrum feeding

Recommendations for colostrum feeding have changed dramatically over the last two decades. Twenty years ago, it was considered acceptable for all calves to run with their dams for 1, 2 or even 3 days and for her to pass on passive immunity through natural suckling. As producers learnt more about the causes and prevention of calf diseases, they became more 'colostrum conscious'. Current advice to farmers is to ensure all calves drink from their dam within the first 3–6 hr of life and, if not, to provide additional colostrum from its mother or another freshly calved cow. Colostrum quality can be assessed visually, or more accurately, using a colostrometer, which works on the same principle as the hydrometer used to measure the acid level in car batteries. Recently, more sensitive field test kits have become available to calf rearers in many countries.

Two feedings during the first day, 6–12 hr apart, and each of 2 L of good-quality colostrum used to be considered sufficient to provide passive immunity, mainly because of concern about the small capacity of the abomasum in newborn calves. However, US advisers now recommend that dairy farmers remove the calf as soon as possible after birth (within 15 min) and feed it 3–4 L of top-quality colostrum at one feeding. This may not be possible with small calves, so it could be spread over two feedings, 6 hr apart. This can be via teat, bucket or stomach tube. The latest findings are that this extra colostrum will be stored in the rumen, from where it slowly passes through the abomasum into the intestines where the Ig are absorbed into the blood.

For those rearers who still wish their calves drink from their dams, the UK guidelines provided by Margerison and Downey (2005) are that suckling should always be supervised, and assisted where necessary, to achieve colostrum intake levels of 2.5 L within the first 6 hr of life and 4 L within the first 12 hr of life. Where the calf is not able to suckle successfully, it should be offered milked or stored colostrum from a nursing bottle or stomach tube and, in these cases, all feeding utensils should be kept scrupulously clean. To ensure adequate quality, the colostrum consumed should be from adult cows,

Table 5.1. Recommended colostrum allowances for newborn calves

Age (days)	Period	Allowance (L)
1	<6 hr	2.5
1	<12 hr	4
2 –3	Daily	3–4
Continued	Continued	5% total milk offered

Where scours is a problem, continued feeding of good-quality colostrum is recommended up to 3 weeks of age.

because heifers tend to have lower immunoglobulin levels in their colostrum. The amount of colostrum required to be consumed according to age is presented in Table 5.1.

This chapter highlights the important principles behind colostrum feeding, to ensure that all calves get a 'good start to life' through adequate transfer of passive immunity. These principles can be categorised into 3 Qs, namely:

- **Quality** is providing good-quality colostrum.
- **Quantity** is ensuring calves ingest sufficient antibodies.
- **Quickly** is timing the first feed to ensure efficient absorption of the antibodies into the blood.

5.1.1 Colostrum quality

Colostrum is produced by the pregnant cow up to 5 weeks before she calves down. If cows are not well managed, colostrum quality could be reduced. Good management includes providing a good-quality diet for dry cows, ensuring they are in good general health and minimising stresses such as climatic or overcrowding during late pregnancy. Older cows, and cows raised in the herd (compared with those purchased as in-calf heifers), will generally produce better quality colostrum, containing more antibodies for those diseases existing on that farm. First-calf heifers are likely to have the lowest levels of antibodies in their colostrum because they have had less exposure to these diseases.

The immune properties of colostrum can be enhanced by vaccinating cows at drying off. There are vaccines to improve calf immunity against *E. coli*, *Clostridia*, *Leptospira* and *Salmonella*, but they may not always be readily available in Asian countries. Selection of the most appropriate vaccines should be based on the prevalence of particular calf diseases in that area – information readily available from local advisers and veterinarians. American farmers are fortunate in having vaccines against two other major causes of calf scours, rotavirus and *Cryptosporidia*, and also against several respiratory infections. As demand for better quality colostrum increases, such dry cow vaccines should become available in Asia.

Newborn calves need to ingest at least 100 g of Ig within their first 3–6 hr of life, and the same amount no less than 12 hr later. The quality of colostrum is expressed in terms of its Ig concentration, with excellent quality colostrum containing at least 90 g/ L, good quality 65–90 g/L, moderate quality 40–65 g/L and poor quality less than 40 g/L. The volume of colostrum that should be drunk to supply 100 g of Ig can then be calculated from its quality.

The higher the colostrum quality, the faster and more efficiently the Ig are absorbed by newborn calves. With poor-quality colostrum, not only must calves be fed very large volumes to provide sufficient Ig intakes, but it is likely that, even then, inadequate amounts of Ig will be absorbed into the blood. For example, 2 L of colostrum containing 80 g/L of Ig will provide more passive immunity than 4 L of colostrum containing only 40 g/L of Ig. Bloody colostrum may also be lower in antibody levels.

After their first milking, dairy cows begin to reabsorb the Ig back into their udder tissue. For this reason, colostrum from the second milking contains only half the Ig content as that from the first milking. Cows are generally deficient in Ig levels if they have been previously milked or are seen to be leaking milk prior to calving. Colostrum quality is also low in induced cows or those with less than 4 weeks between drying off and calving.

On the whole, Jerseys produce colostrum containing more Ig than do Friesians. In fact, very few Friesian cows produce excellent quality colostrum. Cows yielding large volumes of first milking colostrum (8 L or more) are more likely to have low Ig concentrations. There is little seasonal effect on colostrum quality. Some studies have found seasonal differences in acquired immunity in calves, but this is more related to changes in colostrum feeding management rather than colostrum quality.

Recent studies in Australia have confirmed US findings that many dairy cows do not produce good-quality first milking colostrum. This is also likely to be the case with dairy cows in tropical Asia. Using a colostrometer to assess quality, colostrum was categorised (with the percentages of cows) as poor (41%), moderate (34%), good (21%) and excellent (4%). These findings highlight the importance of identifying cows producing poor-quality colostrum soon after parturition then feeding their calves on stored good- or excellent-quality colostrum (Humphris 1998).

Johne's disease (or paratuberculosis) is an ongoing problem in many dairy regions in Western countries, and probably in Asia as well. It is possible for Johne's disease to be transmitted from cow to calf prior to calving, and also via the colostrum. Although the incidence of transfer by these methods may be quite small, it should be taken into account in any Johne's disease prevention program. For this reason, current recommendations are for disposal of both the infected cow and her daughter, if the last calf before breakdown was a heifer. The most important aspect of any Johne's disease control program is to minimise contact between milking cows and replacement heifers until at least 6–12 months of age. Rearing systems based on milk replacers, rather than whole milk, are often recommended, but this is mainly to reduce the possibility of cross-infection from cows being milked in the dairy to calves being fed fresh whole milk in the rearing shed.

The protection from the passive immunity passed onto the calf peaks 1–2 days after effective colostrum transfer and then it declines. By 2 weeks of age, it has declined enough to increase the calf's susceptibility to bacteria, viruses and other pathogens, before the calf's own immunity increases to an effective level. Therefore the calf can be quite vulnerable to pathogen invasions coming from dirty feeding equipment or other sources between 14 and 21 days of age.

In well-managed herds with few disease challenges, calf isolation is particularly important because the colostrum is likely to have lower Ig levels because pregnant cows would have been exposed to fewer pathogens. Providing calves with both isolation and high-quality colostrum is important in these herds, and also those with high culling rates because they would contain a higher proportion of first-calf heifers. Farmers with seasonal-calving herds and practicing oestrus synchronisation of heifers to calve before the older cows face a quandary of not having fresh supplies of high-quality colostrum on hand. These farmers may consider storing colostrum from the previous year, which means they must be able to identify the best quality colostrum for long-term storage. If they are concerned about the quality provided by any freshly calved cows, then additional colostrum should be fed to their calves from a store of good-quality colostrum.

5.1.2 Timeliness of colostrum feeding

Every half hour after birth that colostrum feeding is delayed, antibody transfer decreases by about 5%. A calf that does not drink until 6 hr old has then already lost the opportunity for 30% of the possible antibodies entering its bloodstream.

Margerison and Downey (2005) reported on the mortality of calves suckling colostrum, for the first time, at different hours of life, as shown in Table 5.2.

Colostrum feeding can then be seen as a race between the arrival of the protective colostrum Ig in the calf's intestines and the disease-causing pathogens. The longer calves are without Ig, the more opportunity for these pathogens to invade the gut. If certain pathogens, such as *E coli*, 'win the race' in the first few hours, they can even be absorbed into the blood, causing severe scours and reduce the effectiveness of any absorbed Ig.

The timing of colostrum feeding is crucial. The cells in the intestinal wall mature in these first 12 hr, eventually shutting down their absorption mechanism. Furthermore, after 24 hr, when the abomasum starts to secrete acids to make the milk-digestive enzymes more effective, these degrade the Ig proteins, which reduce their effectiveness. Another confounding factor is that protection against pathogenic bacteria is minimal until the abomasum can secrete sufficient acid to reduce their potential to cause scours. Provided the calf has drunk colostrum, the maternal antibodies can control the spread of these harmful bacteria. The Ig in colostrum is still beneficial to the calf even if it can no longer be absorbed into the blood, as it lines the intestinal walls to provide local protection against the build up of pathogens.

Stressed calves, such as those born in cold, rainy weather and left unprotected, and those requiring assistance during birth, cannot absorb Ig for as long a time as calves with easier births. Calf rearers who feed colostrum by artificial means give these calves a greater chance of survival.

Table 5.2. The mortality of calves suckling colostrum for the first time at different hours of life

Hours post-natal	Mortality (%)
2–6	5
7–12	8
13–24	11
25–48	20

The current US recommendations for first feeding are to offer 3 L to Jerseys and small Friesians and 4 L to average-sized Friesians, as soon as possible after birth. A second feeding of 2 L, or more if the calf is willing, can be provided 6 hr later, although it is not really necessary to feed these calves until the following day. Colostrum and transition milk should then be fed for the first few days of life to provide extra nutrition and local protection against disease.

If calves are not weak from a lengthy and difficult birth and are breathing well, they do not require further stimulation from their dams, such as licking off. There is little benefit for cows to have them lick their calves to strengthen maternal bonds. In searching for the teat, calves are likely to take in more pathogens than they would get from a sterilised stomach tube.

The longer calves spend with their dams, the greater their chances of contracting disease. The practice of 'snatch calving', or removing the calves from their dams at birth, may be difficult to encourage in seasonal calving herds unless there are obvious benefits through reduced health problems, such as Johne's disease, and reduced mortalities. It would greatly increase labour requirements during any busy calving period.

5.1.3 Identifying and storing good-quality colostrum

Over the years, there have been various attempts to assess the quality of colostrum. Unfortunately, visual appraisal is a poor way of assessing quality because thick, creamy colostrum may simply be indicative of its high fat content. There is a negative relationship between Ig level and fat content in colostrum because the Ig resides in the non-fat component of the milk solids. The colostrometer was developed specifically to determine Ig levels, but recent research has shown up some limitations. When using colostrometers to quantify Ig status, it is important to measure it at room temperature, or better still, use a thermometer to make certain the colostrum is measured at the recommended temperature. Colostrometers are good at detecting poor samples, but unfortunately they are limited in their ability to make other assessments. Their major role should be to screen samples to ensure only colostrum with more than 80 g/L of Ig is fed fresh or stored for later use.

Recent research has shown that refractometers calibrated in the brix scale (used to assess sugar content of sugar cane and molasses) can be used to assess the quality of colostrum with good accuracy (Dairy Australia 2011). Its accuracy is not affected by the temperature of colostrum, as is the case with colostrometers.

Colostrum allowed to sour or become overheated loses its antibody effectiveness. Freezing is the best method for storage, because this will retain its antibody activity for at least 1–2 yr. Frost-free freezers are not ideal because their freeze–thaw cycles can allow the colostrum to thaw, and this can shorten its effective storage life. Rapidly chilled colostrum can only be stored in the refrigerator for up to 5 days, before it loses its antibody effectiveness. Containers of stored colostrum should be identified with its origin and, if known, its quality. Colostrum can be pasteurised and stored for up to 10 days, but this is only practical on very large farms. Fresh colostrum should not be mixed with that being stored, so each individual cow's colostrum should be stored in a separate labelled container.

Frozen colostrum should be thawed out carefully, because overheating it to above 50°C denatures the Ig. Colostrum frozen in 1–2 L milk cartons, or better still, in thick plastic bags can be thawed out in 50°C water. The time between the first appearance of the calf's feet prior to birth until it is first ready to drink, should be sufficient for frozen colostrum to thaw in warm water at 50°C. If using a microwave oven, it should have a turning tray to avoid hot spots and the defrost setting should be used. Pour off the liquid as it thaws to reduce overheating.

5.2 Feeding colostrum to newborn calves

Clean the teats of cows from which colostrum is to be harvested and ensure it is collected into clean and sterilised containers. Even moderate levels of bacterial contamination can lead to significant scour problems. Cover the colostrum bucket tightly, both before and after collecting the colostrum, to ensure it remains clean.

US veterinarians now recommend feeding 3–4 L rather than 2 L of good-quality colostrum at first feeding, just to ensure that adequate Ig are ingested. Increasing volumes at first feeding markedly reduces the number of calves with low blood Ig levels, indicative of failure of passive transfer of immunity. It takes the first 2 L to fill the rumen while the second 2 L spills over into the abomasum. Newborn calves should readily drink 2 L through a teat from a nipple bottle, but greater volumes are generally refused. There is little difference in absorption of colostral antibodies whether calves drink it all from a nipple bottle or it has been administered using a stomach tube.

Stomach tubing a newborn calf with its first colostrum.

Stomach tubing is generally necessary to ensure the entire 3–4 L is consumed. Fluids will pass directly into the rumen, not the abomasum, because the oesophageal groove will not close. However, it will quickly flow from the rumen into the abomasum. Farmers should always have a stomach tube on hand and learn how to use it. Calves weak from difficult calvings or with swollen tongues preventing them from sucking should be tube fed the entire 3–4 L of colostrum immediately. Calves will not regurgitate it or get it into their lungs if the fluid is correctly administered with a stomach tube. Veterinarians are not overly concerned about providing the calf with too much colostrum and causing scours, because this first milk contains less lactose than later milk, and it is the overflow of lactose into the hind gut that leads to scours. The first drink is the most important because the ability of calves to absorb further Ig through the gut wall drops off markedly thereafter.

5.2.1 How to stomach tube a calf

Weak calves may not be able to drink liquids from a teat. Stomach tubing is the best way of ensuring that they consume enough liquids. Some dairy farmers routinely stomach tube all newborn calves to provide colostrum and be certain that they will absorb sufficient antibodies to enhance their immunity against diseases. Scouring calves with severe dehydration, which are too weak to drink themselves, can also be stomach tubed.

The stomach tube is a flexible piece of plastic tubing with a tear-shaped end designed to be easily inserted into the oesophagus, but not into the lungs. It is usually attached to a plastic container holding the liquid to be fed.

The first step in using the stomach tube is to determine the length of tube to be inserted. This is measured as the distance from the tip of the calf's nose to the point of its elbow behind the front leg, which is usually 45 cm or more. This point can be marked on the tube with a piece of tape.

Ideally, the calf should be standing so the fluids are less likely to back up and enter its lungs. However, calves that are too weak to stand can be tubed in a sitting position and even while lying down. The stomach tube is easier to use when calves are restrained. Young calves can be backed into a corner for better head control. A calf allowed to throw its head from side to side may injure itself or the operator.

If the weather is cold, the tube can be placed in warm water to make it more pliable. The tube should be dipped into a lubricant, such as mineral or vegetable oil. The tip of the tube is then placed into colostrum or whole milk, whichever is to be fed. Calves may suck the end of the tube, making it easier for it pass into the oesophagus. A calf's mouth can be opened by gently squeezing the corner of the mouth or by holding its head over the bridge of the nose and gently squeezing the upper palate or gums.

Once the mouth is opened, the empty tube should be passed slowly along the tongue to the back of the mouth. When the tube is over the back of the tongue, the calf starts chewing and swallowing it, after which the tube is passed down into the oesophagus. The end of the tube can be felt quite easily in the neck as it passes down the oesophagus. Never force the tube: if it is being correctly put down the oesophagus, it should slide in quite easily.

After the tube is in place, and before any fluids are given, it should be checked for proper positioning in the oesophagus. If it is properly positioned, the rings of the trachea (leading into the lungs) and the rigid enlarged oesophagus can be felt easily. If you cannot feel both of these, remove the tube and start again. The exposed end of the tube should be checked for spurts of air, which indicate that the tube has gone into the lungs. The calf will often, but not always, cough if this occurs.

It is important to ensure that fluid does not leak from the tear-shaped end of the tube while it is being inserted or withdrawn, because this fluid may be inhaled by the calf and cause pneumonia.

The tube can be unclipped or straightened out or the container can be tipped up to allow liquid to flow down into the stomach. Liquids should be at body temperature (38°C) to prevent shock to an already weak calf. It may take 3 min or more to allow sufficient fluid to be administered. The calf will regurgitate less with a slow flow rate.

When feeding is over, the tube should be slowly removed. The tube should be cleaned and sanitised, then allowed to drain and dry.

Veterinarians often concede that this method for colostrum feeding is not natural, but it does provide an easy and well-tolerated method of achieving an adequate and early intake of Ig at the first feeding. It is widely used throughout the world, where calf rearers are finding significant improvements in passive transfer of immunity and reduced calf health problems.

5.2.2 Making best use of transition milk

As mentioned earlier in this chapter, transition milk is that milk produced by the cow after the second milk post-calving. Depending on the policy of the milk processors, this milk should not be commercially sold because it can reduce the efficiency of processing the raw milk destined for dairy products than can pasteurised milk. This milk can still contain some colostral antibodies. As with true colostrum, if fed following the closure of the gut wall to antibodies entering the bloodstream, this milk will still provide protection against pathogens in the calves' stomachs and intestines. It does this by deactivating the pathogens in the gut contents. The greater the problems with scours in the calf shed, the greater the benefit of routinely feeding transition milk, and any excess colostrum, to calves within their first week of life.

Because such transition milk may have no market value, it should be fed to calves to provide nutrients for calf growth. Compared with market milk containing 12–13% milk solids, colostrum comprises 25% and transition milk 15–18% milk solids. There are a lot more nutrients available per L of colostrum and transition milk than in market milk.

Transition milk should be stored in clean containers until fed to calves and can form the initial source of milk before calves are fed CMR, if that is the main form of milk feeding. It can even be blended with CMR powder to reduce the amount of powder required to provide the daily nutrient requirements for milk-fed calves.

5.2.3 Using stomach tubes to relieve abdominal pressure

Provided it is long enough, a stomach tube can also be used to relieve pressure build up in the stomach during milk bloat. This can occur with twice daily feeding of some milk replacers. This is the result of the previous curd of clotted milk not being given sufficient time to digest before the calf is offered further milk replacer.

Milk bloat can also occur when milk enters the rumen, rather than the abomasum, and ferments with other ruminal contents. This can restrict the escape of ruminal gases. This may occur when calves have functional rumens and, in these cases, they can be weaned off milk. Where this problem occurs in immature calves, they should be examined further by a veterinarian.

To relieve pressure build up, introduce the tube, without the bottle attached, into the calf's oesophagus as described above. The gas should then be heard escaping, generally with a foul smell, and the distended abdomen will quickly return to normal.

5.3 Results from overseas research on colostrum feeding

In his survey of calf mortalities on small holder farms in Ethiopia, Jemberu (2004) reported a close relationship between calf mortality and age when calves drink their first colostrum. If later than 6 hr after birth, calves had higher morbidity (signs of sickness) and mortalities than if they consumed it earlier in life. For each hour delay in colostrum feeding in the first 12 hr of life, the chance of a calf becoming ill increases by 10%.

A large-scale survey of 600 dairy farms throughout the US (NAHMS 1994) found more than 40% of newborn calves had immunity levels below those recommended, while 25% of these calves had critically low immunity status. The death rate among all these calves was twice that of calves with adequate passive transfer of immunity. The study concluded that over 20% of the calf deaths could be avoided by ensuring adequate and timely colostrum intakes. Other estimates suggest that 50% of the mortality that occurs in pre-weaned calves is directly related to this inadequate acquisition of passive immunity (Quigley *et al.* 2005).

This study allowed a comparison of the effects of varying blood Ig levels on calf performance for 2020 calves, as presented in Table 5.3. As blood Ig levels increased, calves grew faster, had more efficient feed utilisation, had lower incidences of scours, but, of most importance, had much lower mortality rates. The extremes of mortality were 29% in the 6% of calves with very low blood Ig levels (0–5 mg/mL) compared with only 8% in the 66% of calves with good blood Ig levels (>15 mg/mL).

If calves are left to nurse from their dam for 24 hr, more than 60% do not take in sufficient Ig. In this study, a second group of calves was provided with 2 L of good-quality pooled colostrum as soon as possible after birth, and those calves recorded a 19% failure of passive transfer of immunity. A third group, fed 3 L of colostrum, had a 10% failure of passive transfer.

Herds with high-yielding cows differ in their newborn calf care from those with lower yielding cows. Top producers were more likely to separate calves from dams at birth before nursing, feed colostrum either by bucket or stomach tube, and feed 4 L/calf or more. Bucket feeding the colostrum was more popular than using stomach tubes, although more used stomach tubes in larger herds. Among farmers that allowed cows to nurse their calves, top producers were more likely to supplement colostrum delivery with hand feeding.

As a result of findings such as these, two-thirds of US dairy farmers now artificially feed the first colostrum to their calves. Of those still allowing newborn calves to nurse

Table 5.3. Four-week performance of calves with varying levels of blood Ig levels

Blood Ig level (mg/ml)	0–5	5–10	10–15	15–25	>25
Percentage of calves	6	11	16	29	37
4-week live weight gain (kg)	9.6	10.7	11.0	11.1	11.6
Feed conversion (kg feed/kg gain)	2.7	2.1	2.2	2.0	1.8
Average faecal score[a]	1.4	1.3	1.2	1.2	1.2
Scour days[b]	7.3	5.7	4.8	5.1	4.9
Mortality (%)	29	16	11	8	8

[a]Faecal scores: 1 normal; 2 loose; 3 watery; 4 blood or mucous.
[b]Scour day, any day when faecal score is 2 or more.

from their dam, 40% assist each calf, thereby increasing the calves' chances of receiving adequate and timely quantities of first colostrum. Although providing calves with immediate assistance to stand and suckle increases the ingestion of Ig, many calves still show failure of passive transfer of immunity.

Delayed suckling is the major cause of poor transfer of immunity: surveys have shown that 25–30% of calves fail to suckle by 6 hr and nearly 20% by 18 hr. The first suckling is later in calves born to heifers and in those born to cows with low, pendulous udders. Diseases are more prevalent in calves that have delayed their first suckling.

Furthermore, there is considerable variation in the actual quantities of colostrum drunk by naturally suckled calves, but the average intake is only 2.5 L within their first 24 hr of life. Unless this is very high-quality colostrum, inadequate Ig intakes will occur.

These findings demonstrate several important aspects of colostrum management. The very high failure rate with calves nursing their dams is due to the inability of calves to drink sufficient colostrum within a few hours after birth. The udder of the cow at birth is large, and at times painful, making drinking of sufficient colostrum difficult. Furthermore, calves born weak or having difficult births may not even stand for several hours. As already mentioned, the Ig concentration of the colostrum may be low, meaning that calves have to voluntarily drink large volumes at a time when both cow and calf would prefer to rest.

Because the gut absorbs less Ig following this first drink, it is preferable to prevent the cow from suckling her calf unless the colostrum is guaranteed to be high quality. There will also be room in the calf's stomach for the administration of additional selected colostrum.

Calves fed by teat, bucket or stomach tube absorb Ig with equal efficiency. The presence of its dam during artificial feeding can improve Ig absorption. The colostrum should be warmed to body temperature, so that calves will not require additional body energy during its digestion. This will prevent shivering of calves after drinking cold fluids, which is common in cold and wet weather.

Researchers in Australia have found that the absorption of colostral Ig can vary from 15 to 60% in calves born at the same time. This may be linked to the calves' ability to clot the colostrum in the abomasum. If it fails to clot, it passes to the small intestine where absorption is less efficient.

In this Australian study, 43% of newborn calves failed to form a clot from 2 L of milk within 1.5 hr of feeding. Three factors were found to reduce clotting ability in the abomasum. Firstly, colostrum may not clot as readily as normal milk. Secondly, newborn calves produce rennin of lower clotting ability than that of older calves. Thirdly, amniotic fluid from the dam is usually present in the abomasum, and this can inhibit rennin activity. The addition of rennet was found to improve clotting ability and Ig absorption.

5.3.1 Financial benefits from good colostrum feeding practices

Data from a large-scale US calf-rearing unit allowed for the calculation of the financial benefits arising from optimum colostrum management (Fowler 1999). When comparing performance of 335 calves with low immunity (0–9.9 mg/mL of Ig) to those of 1663 calves with high immunity (>10 mg/mL of Ig), there were four major benefits as follows:

1. **Calf weight gain.** Low-immunity calves gained 10.3 kg compared with 11.3 kg in high-immunity calves. Valuing each kg live weight gain at US$1.50, each high-immunity calf returned at least US$1.50.
2. **Feed conversion.** Low-immunity calves required 2.35 kg feed/kg live weight gain compared with only 1.95 kg feed/kg live weight gain in high-immunity calves. Over 4 weeks, the high-immunity calves consumed 5.4 kg less feed, when valued at US$1.05, representing a saving of US$5.70/calf.
3. **Incidence of scours.** Low-immunity calves had 6.3 scour days compared with only 4.9 scour days in high-immunity calves. Costs of antibiotics and electrolytes were US$10.80 for low-immunity and US$7.10 for high-immunity calves, a difference of US$3.70. These costings did not take into account any additional veterinary bills to treat the scouring calves.
4. **Calf deaths.** Mortality rates were 20.7% in low-immunity calves compared with only 8.6% in high-immunity calves. Valuing each calf at say US$100 each, savings were US$12.10 per high-immunity calf.

The total savings arising from optimum colostrum feeding practices then amounted to US$23, or nearly a quarter the value of each replacement heifer calf. Readers can convert these US values to local currency using the currency converter in Appendix 4.

5.4 Summarising good colostrum feeding management

In summary, the important principles of good colostrum management are:

1. Use colostrum from mature cows that produce less than 8 L at their first milking.
2. Use only first milking colostrum.
3. Feed 4 L to large calves or 3 L to smaller calves at first feeding.
4. Feed colostrum as soon as possible, at least within the first 3 hr after birth.
5. Do not let calves suckle their dams.

A US organisation, the Bovine Alliance on Management and Nutrition (BAMN) has developed a series of farmer guidelines for calf management with one specifically on colostrum (BAMN 1995). Their guidelines for colostrum management are presented in Table 5.4.

Table 5.4. Current recommendations for post-calving management of dairy herds in the United States (BAMN 1995)

Item	Do	Don't
Calving area	Ensure cows calve on a clean calving pad or on clean pasture	Forget to separate springing cows from the herd or forget to clean pens between cows
Separating calves	Separate calf from dam as soon as possible	Leave calf with dam for more than 1 hr
Colostrum collection	Clean cow's udder and teats before milking Milk out cows as soon as possible after birth (within 1–2 hr) Collect colostrum in clean, dry container	Collect colostrum from a dirty udder Wait 6 hr or more before milking Mix colostrum with transition or hospital milk
Colostrum quality	Measure colostrum quality before feeding Use only good-quality colostrum Use fair- or poor-quality colostrum and transition milk only as nutrition for older calves	Feed colostrum from cows tested positive for Johne's disease, leucosis or mycoplasma Use thin, watery colostrum, especially from heifers Use bloody or otherwise abnormal colostrum Feed poor-quality colostrum at the first two feedings
Colostrum storage and handling	Chill colostrum immediately after collection Refrigerate colostrum at 4°C for less than 24 hr Freeze good-quality colostrum in 1 or 2 L containers Thaw colostrum carefully to preserve antibodies	Pool colostrum Store colostrum at room temperature Store frozen colostrum longer than 1 yr Thaw colostrum in extremely hot water or in a microwave under high power for >1 min at a time (as this destroys antibodies)
Colostrum feeding	Feed first feeding of colostrum as soon as possible (within 1 hr) Use fresh colostrum from the dam, if good quality Use one of the following options to feed colostrum: A. Feed 2–3 L in the first feeding and again within 8 hr B. Feed 4 L via oesophageal feeder within 1 hr of birth Use an oesophageal feeder regularly if the calf will not consume sufficient colostrum; clean and disinfect feeder between calves	Use colostrum from cows that are leaking colostrum prior to or at calving Use bloody or mastitic colostrum Wait for the calf to stand and nurse Allow the calf to nurse the dam Feed less than 2 L per feeding Use a broken or dirty oesophageal feeder Use the same oesophageal feeder for sick calves and for feeding colostrum
Other management tasks	Dip navels in or spray them with tincture of iodine as soon as possible Put calf in an isolated, dry and draft-free environment Continue to feed lower-quality colostrum or transition milk for 2 or 3 days after birth	

Farmers who do not practice good colostrum feeding management may find their calves still remain healthy and grow well. Certainly, some calves with low blood Ig levels are healthy and productive. This reflects the importance of other aspects of calf rearing, such as good hygiene, lack of climatic stress, empathy from rearers and good management of milk feeding. However, as our expectations for good pre-weaning calf performance increase, the importance of improving immunity against disease becomes paramount. It is highly unlikely that dairy farmers can develop a calf-rearing system with minimal health problems and that routinely produces weaned heifer replacements weighing 100 kg at 12 weeks of age without a sound colostrum feeding program.

6

Calf and heifer mortalities in the tropics

This chapter documents the high calf and heifer mortalities reported in many tropical dairy farming surveys.

The main points in this chapter

- Tropical areas are not ideal locations for calf rearing because the high temperatures and humidities introduce many potential disease problems to young dairy stock.
- In addition, the type of dairy farming (generally poorly resourced small holder farming) and the general lack of awareness of the long-term implications of poorly reared stock do not encourage farmers to pay close attention to their calf- and heifer-rearing systems.
- Surveys of calf-rearing systems in tropical Asia, Africa and South America highlight the high calf and heifer mortalities.
- Pre-weaning calf mortality rates of 15–25% would be typical on many tropical dairy farms, but they can often be as high as 50%, indicating very poor calf management.
- This contrasts with 'gold standards' in the US of less than 8% mortality from birth to 6 months, while Australian farmers only suffer 3% losses.
- The involvement of farmers in simple extension programs in Sri Lanka and Kenya have drastically reduced calf mortalities and improved pre-weaning growth rates.
- Problems in calf rearing, as identified on Tanzanian small holder farms, have been ranked in decreasing order by farmers as labour, poor calf growth, diseases, little milk available and inadequate knowledge, which all led to high calf mortalities.
- Clearly, calf mortality in tropical SHD systems is a bigger problem than in temperate farms, for a variety of reasons.
- Unfortunately, such high rates are accepted as 'normal' on many farms, whereas they could be dramatically reduced by following a few relatively simple procedures.
- The research has yet to be undertaken (or at least reported) in the tropics where young stock are reared using the best management practices found on temperate farms, to ascertain more realistic targets for tropical dairy small holdings.

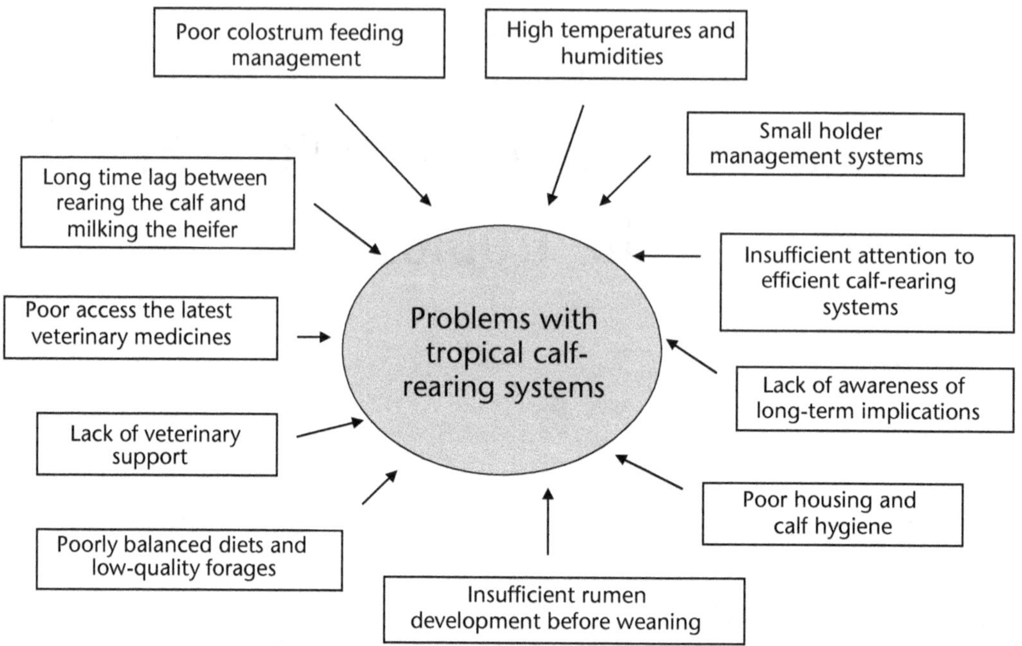

Figure 6.1. The major problems with many calf-rearing systems in the tropics

6.1 Difficulties with rearing replacement heifers in the tropics

Tropical areas are not ideal locations to rear young dairy stock for many reasons, such as those highlighted in Figure 6.1. The harsh tropical climate introduces many problems to milk-fed calves, while the type of dairy farming (generally poorly resourced small holder farming) and the lack of awareness of the long-term implications of poorly reared stock, do not encourage farmers to pay close attention to their calf-rearing systems.

The reasons for poorly reared calves are many and include:

- The tropical environment encourages the proliferation of many disease organisms that can reduce calf and heifer performance. These include some diseases that only occur in the tropics.
- Dairy cows are essentially temperate animals, which are most comfortable at 6–18°C. The high tropical temperatures and humidities introduce specific climatic stresses that adversely affect calf and heifer feed intakes, growth rates and fertility.
- As ambient temperatures approach body temperatures, stock rely more on cutaneous evaporation to remove residual body heat, and this is less efficient than other forms of heat loss, requiring increasing amounts of water and minerals.
- The stock on many small holdings are multi-purpose, being farmed for manure, meat and even draught, in addition to milk. Young stock management is often not a high priority on such farms.
- Many of the dairy farms are small holdings where farmers often lack the resources to develop the most effective rearing systems for young stock, because most of their attention is directed towards income generation; namely feeding and managing their milking cows.

- The best way to reduce calf mortality is to practice good husbandry. Some common ailments can be treated by the farmer, but, once a calf is seriously ill or large numbers are ill, expert veterinary advice should be sought, but is not always readily available.
- The levels of shed hygiene are often sub-optimal and this can have detrimental effects on monogastric animals, such as milk-fed calves.
- The nutritive value of tropical forages is generally poorer than that in temperate areas, so weaned heifers cannot grow as fast as they do in temperate regions without continual access to high-energy and protein supplements. Because these are generally more expensive than forages, farmers are less likely to provide sufficient amounts.
- Sub-optimal feeding regimes, leading to low growth rates, can greatly reduce feed efficiency in replacement heifers.
- In many cases, farmer extension programs place little emphasis on feeding and managing calves and heifers because of the extended time required for such investments to reap rewards.
- Service providers and agribusinesses are less able to source the most up-to-date equipment and farm inputs, such as the latest generations of veterinary drugs or calf milk replacers that are more readily available to farmers in the developed temperate dairy regions.
- Many veterinarians are not fully aware of the most recent treatment and prevention animal health protocols readily adopted by dairy farmers in more developed countries.
- Many farmers are not fully aware of the high costs associated with poor management of young stock, arising through, firstly, the high wastage rates of calves and heifers and, secondly, the detrimental effects on potential milk yields and fertility.
- Delayed calving can greatly increase the total costs of rearing replacement heifers.
- We know less about the constraints to the performance of young stock in the tropics because there has been less research undertaken. Accordingly, there is less relevant extension material available for dissemination to tropical dairy industries.

The 'bottom line' is that fewer resources are devoted to young stock management in the tropics.

With increasing numbers of temperate, poorly adapted dairy stock being imported into tropical Asia, there is an even greater need for well-planned and conducted farmer training programs in these regions.

Considering all the above, it is not surprising that the performance of young stock on most tropical, or even subtropical, dairy farms is below that observed on temperate farms. Not only is the management of young stock sub-optimal, such documentation is often difficult to find, particularly from Asia.

6.2 Surveys of calf and heifer mortalities in the tropics

The results of recent published surveys on mortalities during calf rearing undertaken in tropical dairying areas are presented in Table 6.1. Altogether there are 17 studies reported, mostly in Africa and South America, but others in India, Sri Lanka and Vietnam. The surveys covered a wide range of dairy production systems, from large-scale grazing and feedlots to small holder systems. The highest pre-weaning mortality

Table 6.1. Details of studies documenting calf mortalities in tropical dairy farming areas, listed in decreasing level of pre-weaning calf mortality

Location and year	Farms surveyed	Mortality	Further details	Reference
1. India 2005?	90 farms; 30 with 1–10 cows, 30 with 11–20 cows and 30 with >20 cows Restricted suckling Data to weaning?	68% in small herds 82% in medium herds 84% in large herds **81% pre-weaning**	Only 22% of farms reared dairy cattle calves Only 33% of farms reared buffalo calves Important causes are scours, endo- and ectoparasites, navel ill and pneumonia Only 17% of farms dewormed calves Only 17% of farmers call veterinarian if calf falls sick Farmers consider calf rearing uneconomic so only purchased mature stock	Tiwari *et al.* (2007)
2. Zimbabwe 1996	50 small holder farms 261 dairy stock, various sire breeds	4.7% stillbirths **35.0% mortality to 12 months**	Female mortalities: 25.3% to 12 months, 6.1% from 1 to 2 yr, 11.4% from 2 to 5 yr Jerseys had higher and Red Danes had lower calf mortalities	French *et al.* (2001)
3. Ethiopia 1997–1999	Five large farms with Friesian and Jersey cows 701 calves	19.7% in first year 4.8% from 12 to 18 months 5.2% from 18 to 24 months **29.7% from 0 to 24 months**	Higher in Friesians than Jerseys High loss 0–90 days and 5–6 months (post-weaning) First service at 28 months (22 months Jersey, 30 months Friesian) CCI: 6.4 months Friesian, 4.8 months Jersey	Asseged and Birhanu (2004)
4. Sri Lanka 1999	Only documenting 'control' or non-assisted farms 25 coconut grazing (CG) and 23 peri-urban (PU) farms 340 dairy XB cows	50–60% of the mortalities before 3 months **23.4% (CG farms) pre-weaning** **28.7% (PU farms) pre-weaning**	First lactation peak milk yield; 4.0 (CG) versus 5.9 (PU) kg/day Calves weaned at 5.9 (CG) versus 3.3 (PU) months 24 month weight: 233 kg (CG) versus 221 (PU) kg AFC: 36 (CG) versus 38 (PU) months	Nettisinghe *et al.* (2004)

Location and year	Farms surveyed	Mortality	Further details	Reference
5. Venezuela 1977–1979	One commercial farm with Friesian and Brown Swiss cows 1656 observations	2.5% stillbirths 15.6% deaths and culls to 9 months 28% in first week, 64% in first month, 18% in second month and 17% from 3 to 9 months and 9.9% from 9 months to first calving 94% of losses due to deaths **25.5% lost up to first calving**	Major causes are scours, pneumonia and joint-ill Losses up to first calving: 32% in Brown Swiss and 25% in Friesians	Vaccaro and Vaccaro (1981)
6. Tanzania 2002	125 small holder farms, stall feeding 977 dairy cattle	10% pre-weaning mortality **25% annual calf mortality**	Average milk yield, 8 L/day AFC: 33 months CI: 20 months	Kivaria *et al.* (2006)
7. Kenya 1991–1992	78 farms with 201 calves European × local dairy stock Weaned at 12 weeks of age	26.6% morbidity up to 12 months **21.6% mortality up to 12 months**	Scours was major cause of death Older calves had higher morbidity Poor shed hygiene related to higher mortality Clinical illness, low red blood cell count and concentrate feeding associated with higher mortality Mineral feeding associated with lower mortality	Gitau *et al.* (1994)
8. Tanzania 1999–2003	Two villages with 269 calves 25% Friesian × 75% Boran	**20.8% calf mortality**	12 months live weight only 88 kg 233 kg at 21 months (first conception) 4 kg/day milk yield during first lactation	Msanga and Bee (2006)
9. Ivory Coast 1997–1999	Three herds of 165 N'Dama cattle with day-time grazing and night-time yarding Limited milking for home consumption	**19% mortality within 7 months**	200–250 kg mature weight 250–600 kg milk/lactation 52% annual calving rate	Knopf *et al.* (2004)

Location and year	Farms surveyed	Mortality	Further details	Reference
10. Kenya 1990–2003	Four large-scale farms with Friesian XB cows 3508 calves	**19% mortality before first calving** Another 6% culled before first calving 22% mortality before 48 months Another 12% culled before 48 months	Most deaths within first 12 months AFC 35 months with range 30–45 months Total deaths and culls prior to AFC increased from 5% to 45% over 13 yr of study	Menjo *et al.* (2009)
11. Ethiopia 2003–2004	Three large and 112 small holder farms 236 Friesian cross calves up to 6 months old	61.5% morbidity **18.0% mortality**	Scours was major cause of mortalities Mortality related to calf age, age at first colostrum and shed cleanliness	Jemberu (2004)
12. Ethiopia 2003?	300 small holder farms Zebu × dairy breeds Calves suckled or bucket fed	1.4% abortions **17.4% pre-weaning**	CCI: 7.8 months	Lobago *et al.* (2006)
13. Mali 2002–2004	38 peri-urban dairy herds Zebus and XB dairy breeds 762 calves	3.9% in first 10 days 5.4% from 10 days to 3 months 4.2% from 3 to 6 months 3.4% from 6 to 15 months **16.9% from 0 to 15 months**	Higher in modern and large farms compared with traditional farms Higher in rainy season Important causes were peri-natal, accidents, digestive disorders (including parasites) and starvation	Wymann (2005)
14. Kenya 1996–1998	Five large-scale farms 673 Friesian calves Data to weaning unknown	**15.6% mortality**	Higher in heifers than bull calves Higher in calves born Jul–Sep Highest in 2–4.5-yr-old dams Pneumonia most important cause Higher in moveable compared with permanent pens	Bebe *et al.* (2001)
15. Brazil 1977–1981	One farm 614 Friesian × Guzera heifer calves Six breed groups Bucket reared to 4 months	Mortality to 12 months 10.1–20.4% range for six breed groups **15.5% average mortality**	Scours, respiratory and tick-borne diseases were major causes of mortality Intermediate crosses had lower mortalities	Madalena *et al.* (1996)

Location and year	Farms surveyed	Mortality	Further details	Reference
16. Ethiopia 1983–2003	One farm 1829 Fogera (Zebu) × Friesian calves Suckled dams until 6–8 months of age	Mortality rates: 3.6% to 30 days, 5.8% to 180 days, 6.5% to 240 days **9.7% mortality to 12 months**	–	Amuamuta *et al.* (2006)
17. Vietnam 2002–2003	99 small holder dairy farms (two to three Zebu XB cows/farm) Data to weaning unknown	3.1% per annum abortions 2.3% per annum peri-natal mortality (first 24 hr) 4.7% per annum calf mortality **7% per annum total calf mortality**	3500 L milk/lactation, 2.5 month dry period, 13.0 months ICI Cows calved twice and 60 m old on average	Suzuki (2005)
18. US 2009	'Gold standards' for calf rearing in temperate regions	Mortality: <5% from 1 to 60 days <2% from 2 to 4 months <1% from 4 to 6 months **<8% from 0 to 6 months**	Morbidity for scours (S) and pneumonia (P): 1–60 days; <25% (S), <10% (P) 2–4 months; <2% (S), <15% (P) 4–6 months; <1% (S), <2% (P) **0–6 months; <28% (S), 27% (P)**	DCHA (2009)

Abbreviations: XB = crossbred; CCI = calving to conception interval; AFC = age at first-calving; ICI = inter-calving interval

Heifer mortalities are very high in overcrowded, poorly cleaned pens.

reported was 81%, from a survey of various Indian dairy systems (Tiwari *et al.* 2007) in which it was found that over half the 90 farmers ceased to rear any calves because they considered it uneconomic and it was cheaper for them to purchase all their herd replacements. This was an extreme finding because the calf mortalities in the other reported studies varied mainly from 10% to 35%, although one Vietnamese study recording only 7% calf mortality.

Unfortunately, the studies documented a wide variety of ages over which calf mortalities were reported, making it difficult to fully interpret the data. However, a range of 15–25% pre-weaning and early post-weaning mortality rates would be typical on many tropical dairy farms. De Jong (1996) highlighted these problems when he concluded that typical calf mortalities in the tropics of 20–45% are much higher than the 7–16% found in temperate regions, because of higher disease incidence and low feeding levels. This was further compounded by the low value of calves reared for meat or replacement heifers. With regard to its potential economic impact, Amuamutu *et al.* (2006) calculated that a calf mortality of 20% can reduce dairy farm net profits by 38%, compared with a target mortality level of 5%.

For comparison, DCHA (2009) present 'gold standards' for dairy farms in the US: namely less than 8% mortality from birth to 6 months (with less than 28% of the calves

Weaned heifers require better management than this.

suffering from scours or pneumonia). 'Gold standards' for post-weaning mortalities are less than 1% from 6 to 12 months and less than 0.5% from 12 months to first calving. Accepted norms are even lower in Australia, where a recent survey of 300 dairy farmers reported mortalities due to illness in calves reared to weaning to be only 3.3% (McNeil 2009).

In his survey of 112 SHD farms in Ethiopia, Jemberu (2004) reported that many of the farmers considered high calf mortality to be their number one herd health problem, compared with mastitis and infertility. On three larger farms he surveyed, calf health problems were generally rated second behind mastitis. The incidence of calf health morbidity (or disease incidence) was higher on the large farms, probably due to the greater number of calves being reared, hence the potential for transfer of pathogens. However, despite the lower morbidity on small holder farms, mortality rates were higher: an observation that the author interpreted as poorer management of sick calves and lack of easy access to health professionals for calf treatment.

Calf mortality in the tropics is undoubtedly very high, often as high as 50% but that almost invariably denotes poor calf management. The best way to reduce calf mortality is to practice good husbandry, the topic of this manual. Some common ailments can be treated by the farmer, but if large numbers of calves show disease symptoms or if stock become seriously ill, expert veterinary advice should be sought.

Table 6.2. Wastage rates for imported and local/mixed stock

Wastage from	Imported stock	Local/mixed stock
Abortions (%)	12.1	6.7
Stillbirths (%)	9.2	6.0
Pre-weaning deaths (%)	37.6	19.9
Pre-weaning culls (%)	1.4	1.9

Source: Vaccaro (1990)

6.2.1 Poor adaptation of imported stock

High calf mortality can also be the result of poor adaptation of exotic stock, such as when unacclimatised temperate or crossbred dairy stock are imported to the tropics, thus adding severe climatic stress to the other hazards of calf life. Vaccaro (1990) reviewed published data on the losses of European dairy breeds in the tropics, compared with European × Zebu crossbreds. Approximately 15% of the imported calves failed to reach first calving or survive their first year of life. He separated data for European breeds as to whether the dams of the calves stock were imported or they originated from local or mixed origin, reporting the wastage rates shown in Table 6.2.

Following weaning, the average wastage rates up to first calving for both groups amounted to 3.9% heifer deaths and 6.6% heifer culls. Over their herd life, imported stock only produced 2.6 calves compared with 3.1 for the local cows. For every 100 females conceived, an average of 56 would have died or been culled before first calving in the case of imported dams, compared with 39 for local dams.

Clearly, young stock management of exotic dairy stock or their progeny require closer attention than was the case prior to 1990, and possibly even after. Anecdotal stories during the 2000s of high mortalities among imported European dairy heifers, milking cows and their progeny in various Asian countries clearly indicate that their survival, as well as their post-calving performance, could still be vastly improved.

6.2.2 The effect of grazing system on calf and heifer mortalities

Table 6.3 presents the results of a detailed survey, undertaken between 1996 and 1998, in which Bebe (2008) and Bebe *et al.* (2003) documented the dairy herd dynamics of 1755 small holder farms in the highlands of Kenya. Of these farms, 987 had dairying enterprises as part of their mixed farming systems, and these had evolved to diversify the risks from dependence on a single crop or livestock enterprise. There was a large movement of weaned heifers from their home farms to other dairy farms as purchased heifer replacements. This strategy was used by farmers to reduce spending scarce feed resources on raising, as yet, unproductive stock.

In the densely populated highlands of Kenya, farmers have had to intensify their dairy systems by changing from free grazing to zero grazing, due to decreasing farm sizes. This led to changes in herd structure, with more emphasis on milk production and increased stocking rates. High stocking rates can be maintained through cut and carry feeding of Napier grass and crop residues, as well as sourcing additional fodder from neighbours and communal areas, together with purchasing more concentrates.

Table 6.3. The effect of grazing system on heifer rearing and herd dynamics on 987 farms in the Kenyan highlands (Bebe 2008)

Grazing system[a]	Free grazing	Semi zero grazing	Zero grazing	Average	Sig[b]
Number of farms	227	326	434	–	–
Farm size (ha)	2.4	1.8	0.9	**1.7**	*
Herd size (number)	4.3	3.1	2.1	**3.2**	*
Cows (number)	2.2	1.7	1.2	**1.7**	*
Calves fed milk using buckets (%)	38	61	68	**60**	*
Milk fed to calves to 3 months of age (L/calf)	321	230	193	–	*
Average weaning age (months)	5.5	4.6	3.8	–	*
Proportion of heifers to cows (%)	40	36	29	–	*
Cow milk yield (L/day)	3.8	4.8	5.6	–	*
Stocking rates (TLU/ha)	1.1	1.0	1.4	**1.2**	*
Proportion of cows in the herd (%)	51	55	62	**53**	*
Annual calving rate (%)	69	51	52	**58**	*
Age at first calving (yr)	2.8	2.7	2.5	**2.7**	*
Heifer calf mortality (%)	15	13	15	**14**	–
Heifer mortality (%)	8	12	7	**9**	–
Cow mortality (%)	13	14	12	**13**	–
Bull calf mortality (%)	21	19	14	**18**	–
Immature bull mortality (%)	16	16	11	**14**	–
Mature bull mortality (%)	13	10	13	**12**	–
Heifers dead before reaching breeding age (%)	27	31	25	**28**	–
Heifers sold before reaching breeding age (%)	11	15	22	**15**	*
Heifers reaching breeding age (%)	62	54	53	**57**	*

[a]Classification of stock. Heifer calf, pre-weaned female; Heifer, post-weaned until first calving; Cow, after first calving; Bull calf, pre-weaned male; Immature bull, post-weaned to 3 yr old; Bull, older than 3 yr.

[b]Sig: An asterisk (*) indicates a statistically significant difference between grazing systems.

TLU = Total Livestock Unit; Stocking rate weighted for bulls (1.0), cows (0.7), heifers/young bulls (0.5) and calves (0.2).

Farm sizes and total herd numbers decreased as farms became more intensive. In addition, calves were more likely to be bucket fed, they were fed less milk to 3 months of age, and were weaned earlier. The cows produced more milk. However, the proportion of heifers to cows was reduced because they were younger when sold to other farmers as heifer replacements. To provide sufficient heifers for a more intensive dairy industry, Bebe (2008) recommended an increasing need for heifer breeding and growing farms.

Mortality rates of all classes of stock were very high, varying from 7 to 19%, but were not statistically significant across grazing systems for any of the animal classes. Losses due to animal diseases and their interaction with nutritional level were a major constraint to small holder dairying, irrespective of the grazing system. With 25% of the heifers dying and 22% being sold before breeding, only 53% of the heifers on the zero grazed farms could have become pregnant, produced milk and hence generate farm income.

The high mortality and low calving rates then led to very high reproductive wastages, irrespective of the level of farm intensification. Together with the numbers of heifers sold, many of these farms were not able to maintain sufficient heifers for their own herd replacements. Herd sizes could not then increase without an external supply of replacement animals. Constraints to rearing replacement heifers on these farms ranged from high losses due to disease and malnutrition through to inadequate breeding services and credit.

6.3 Impact of extension programs on calf and heifer performance in the tropics

Clearly there is much to improve on most tropical dairy farms (whether large scale or small holder) with regards to calf rearing. Two studies have quantified the impact of simple extension programs on the improved performance of milk-fed calves and weaned heifers in the tropics.

6.3.1 Sri Lankan study

To reduce calf mortality and age at first calving of heifers in an artificial insemination heifer calf-rearing scheme during the 1990s, the Sri Lankan Ministry of Livestock Development instigated a heifer-rearing program that supported small holders with subsidised calf meal together with free mineral mixtures, drugs and acaricides for 30 months. Nettisinghe *et al.* (2004) reported the outcome among 100 small holder farmers in the Coconut Triangle grazing and the peri-urban dairy production systems in the wet lowlands of Western Province. About 50 scheme and 50 non-scheme farmers were surveyed in the two production systems 3 yr after the scheme finished and the data are summarised in Table 6.4. Virtually all the coconut grazers were part-time dairy farmers and supplied the formal milk market (receiving 10–11 Rs/L in 1999), whereas the peri-urban farmers supplied the informal milk market (receiving 19–20 Rs/L), while about one-third of them were full-time dairy farmers. About half the coconut grazing farmers stall fed their stock whereas most of the peri-urban farmers employed grazing only.

Live weight at 24 months varied between breeds with Friesians being the heaviest (244–251 kg), followed by Australian Milking Zebus (AMZ, 219–241 kg), then Jerseys (220–248 kg), with the lightest being Sahiwals (191–204 kg). Scheme heifers calved five months earlier than non-scheme heifers (32 versus 37 months) while coconut grazing herds calved 3 months earlier than peri-urban herds (33 versus 36 months). Across breeds, scheme Jersey and AMZ heifers were the youngest to calve (28 months) with non-scheme Sahiwal being the oldest (42 months).

Calf mortalities were lower in the coconut grazing compared with the peri-urban herds (18 versus 21%) and on former scheme compared with the non-scheme farms (13 versus 26%), with 50–60% of these mortalities occurring in the first 3 months of age. In coconut grazing herds, scheme farmers had more young female stock present per cow, whereas there was no difference with the peri-urban farmers.

Table 6.4. The benefits in belonging to an AI heifer-rearing scheme in the Western Province of Sri Lanka (Nettisinghe *et al.* 2004)

	Coconut grazing		Peri-urban	
Production system	**Scheme**	**Non-scheme**	**Scheme**	**Non-scheme**
Number of cattle/cows per farm	5.3/2.5	5.7/2.8	8.3/4.4	8.5/4.4
Peak milk yield in first lactation (kg/day)	4.2	4.0	5.8	5.9
Concentrate fed at peak yield (kg/day)	0.8	0.6	2.0	2.1
Calf weaning age (months)	5.7	5.9	3.3	3.2
Weight at 24 months (kg)	273	233	238	221
Age at first calving (m)	31.1	36.0	33.5	37.9
Calf mortality (%)	12.5	23.4	14.2	28.7
Young female stock present per cow	0.9	0.6	0.6	0.5
Deworming frequency	2.6	1.5	1.4	1.4
De-ticking frequency	5.1	3.7	4.8	4.1
Minerals fed (kg)	2.0	1.2	1.0	1.5
Milk fed (L/calf)	660	572	328	298
Concentrates fed (kg/calf)	646	112	569	66
Concentrate costs (Rs/calf)[a]	847	492	577	518
Total rearing costs (Rs/calf)	14737	6432	14419	6531
Total returns (Rs/calf)	22598	12666	27482	12434
Total profit (Rs/calf)	7861	6234	13063	5903

[a]Rs = Sri Lankan rupee.

Most non-scheme and former scheme farmers applied deworming drugs and fed concentrates, whereas less than half applied de-ticking treatments, while very few fed minerals. On the coconut grazing farms, former scheme farmers dewormed their calves more frequently and spent more money on concentrate feeding than did non-scheme farmers. There was no difference in these calf management practices in former scheme or non-scheme peri-urban farmers. The former scheme peri-urban farmers had higher costs and, because of their higher unit milk returns, they still achieved higher profits than did the coconut grazing farmers.

Poorer calf housing and low level of health management led to peri-urban farmers' herds experiencing higher calf mortalities (see Table 6.4 for details). The coconut grazing farmers' herds produced more female stock per cow because of reduced calf mortality and younger ages at first calving. Because the peri-urban farmers were less motivated to continue the improved calf-rearing practices after the scheme was stopped, differences between former scheme and non-scheme farmers were much smaller.

The total production cost per in-calf heifer was considerably cheaper than the production of such animals, either by multiplication on state farms or imports. The authors calculated the cost of importing one exotic heifer to be the same as producing 16 well-adapted heifers from this study. As in other countries, importation of exotic stock or multiplication of breeding stock on state farms has a poor record for expanding dairy populations. This study showed that only in certain production systems – in this case grazing under coconuts – were farmers likely to continue their improved heifer management practices after such government-supported schemes had finished.

An additional observation from this study was the very high levels of milk fed to rear heifers for the coconut grazing system compared with the peri-urban farmers (600 versus 300 L/calf). For a cow producing, say, 2500 L in a lactation, that represents 12% of her milk fed to rear a calf.

6.3.2 Kenyan study

Lanyasunya *et al.* (2006) reported on a study of 120 small holder farmers in Kenya. It involved 60 test farmers, each provided with a chest girth tape (to estimate live weight), a spring balance and 10 L bucket (to weigh feed), a graduated jug (to measure milk) and a note book. An additional 60 control farmers continued with their traditional calf-rearing practices. Each of the test farmers was also provided with moveable calf pens, compartmentalised with feed and water troughs, a daily heifer feeding schedule and forage planting material (which included improved tropical grasses and legumes). Test farmers regularly dewormed their calves and were visited every fortnight by extension staff. The dairy production systems were categorised as zero grazing, semi zero grazing or free grazing.

Pre-weaning growth rates were higher on the test farms (0.37 versus 0.31 kg/calf/day) while calves offered legume forages (lucerne, *Desmodium and Leucaena)* grew faster than those only offered sweet potato vines (0.40 versus 0.34 kg/calf/day). Mortality of heifer calves on zero grazed farms was lower than on semi zero and free grazed farms (6 versus 15 versus 20%). Mortality of bull calves was higher than for heifer calves, at 13 versus 6% on test zero grazed farms and 11 versus 9% on control zero grazed farms. Heifer live weight at 24 months was 295 kg for zero grazed and 296 kg for semi zero grazed compared with only 240 kg for free grazed stock.

The key messages from this study included:

- The beneficial impact of legumes enabled calves to use low-quality forages more efficiently.
- Calf survival was improved because individual calf pens reduced disease transfer.
- Better performance of zero grazed stock was due to greater attention and improved management.

6.4 Results from other calf and heifer studies in the tropics

6.4.1 Tanzanian study

Lyimo *et al.* (2004) surveyed calf-rearing practices on 60 small holder mixed farmers in Tanzania, with average herd sizes of only three cows. The area had fertile soils and a favourable climate, therefore cropping was given greater emphasis than milk production. Most farmers practiced restricted suckling. Calf survival and growth rates were low, but no mortality data were provided. Calves weighed only 96 kg at 6 months of age. Calves stayed with their dams until 4–7 days, were only provided with concentrates after 4 weeks and were weaned at 4–6 months of age. Calves were then stall fed with limited forages and poorly formulated concentrates. The major constraints to forage supplies were season of year, labour, ease of harvesting and distance to forage supplies. Farmers ranked the problems in calf rearing (mean score out of 8) as follows:

- labour (7.5)
- poor calf growth (7.0)
- diseases (6.5)
- little milk available (6.5)
- inadequate knowledge (4.8)
- mortality (3.4).

They ranked the options for improvement (mean score out of 8) as follows:

- strategic feeding (7.7)
- provide more supervision and adequate management (6.3)
- use cheaper concentrates (5.5)
- farmer's training (5.5)
- adopt more appropriate housing (5.3)
- allocate more funds to calf rearing (5.0)
- accept technical advice (3.8).

It was not easy to say what 'disease' meant to these Tanzanian farmers. In a general sense, it meant anything that was not 'normal'. Disease was then a condition which was detrimental to the health and wellbeing of that animal. It included injuries, infections by micro-organisms, infestations of parasites, nutritional deficiencies, poisoning and hereditary abnormalities. The presence of disease, acute or chronic, reduced cow performance, which invariably reduced farm production and profitability.

Many of their problems were nutritional, with strategic feeding of milk, concentrates and forages seen as most important. These can be solved by improved farmer training and supervision to practice more appropriate husbandry.

6.4.2 Philippines study

A study was undertaken in the Philippines in which heifers were reared in different systems from 3 to 12 months of age and their performance monitored (Payne *et al.* 1967). They were raised indoors or at pasture, or a combination of the two, with or without strategic drenching for internal parasites (Table 6.5).

The indoor/outdoor system was used as it provided protection from heat stress during the day while at night the larvae of most internal parasites would move down the stems of the pasture so calves could safely select the better quality components of the forages. Calf growth rates were highest (0.40 kg/day) in the two drenched groups with access to grazed pasture (outdoor and indoor/outdoor), while mortalities were lowest in

Table 6.5. Performance of Friesian cross calves from 3 to 12 months of age under various rearing systems (Payne *et al.* 1967)

Management system	Drenched	Growth rate (kg/day)	Mortality (%)
Indoor (day and night)	+	0.38	6.7
	–	0.30	0
	Mean	**0.34**	**3.3**
Indoor (day) – outdoor (night)	+	0.40	10.0
	–	0.38	7.7
	Mean	**0.39**	**8.8**
Outdoor (day and night)	+	0.40	16.7
	–	0.37	40.0
	Mean	**0.38**	**28.3**

the indoor groups. Even allowing for their higher mortality, the overall production was superior in the indoor/outdoor and drenched group. The most dramatic finding was the very high mortality in the outdoor and undrenched group.

6.4.3 Conclusions

Clearly, calf mortality in tropical SHD systems is a bigger problem than in temperate farms, for a variety of reasons. Unfortunately, such high rates are accepted as 'normal' on many farms, whereas they could be dramatically reduced by following a few relatively simple procedures. It is quite likely that the DCHA (2009) 'gold standards' in the US may not be achievable in tropical small holder environments. The research has yet to be undertaken (or at least reported) in the tropics where calf and heifer rearing follows the best management practices found on temperate farms, to ascertain more realistic targets for tropical dairy small holdings. The key procedures that clearly require closer attention include:

- ensuring newborn calves ingest sufficient good-quality colostrum and absorb sufficient immunoglobulins to provide adequate passive immunity against calfhood diseases
- providing calves with a hygienic environment, both in the rearing pen and with clean and frequently sterilised feeding utensils
- following a pre-planned feeding program to encourage both early rumen development and adequate growth rates
- following an animal health protocol to minimise the adverse effects of exposure to local disease pathogens.

7

Facilities for calf and heifer rearing in the tropics

This chapter discusses the housing of young stock and other facilities required in a well-designed calf shed.

The main points in this chapter

- Close attention should be given to the design, construction and maintenance of facilities for calves and heifers.
- Considerations are to optimise feeding management, stock welfare and wellbeing, climate control, hygiene and disease management.
- Stock should always be provided with adequate feed and water.
- Herd management practices should always minimise stress, injury and disease.
- Staff should be adequately trained or experienced in good animal care.
- Calves can be housed in individual pens, providing 1–2 m^2 of floor space with a well-drained floor.
- Containing each calf in a raised metal cage provides the best housing because they are isolated from each other, live in a well-ventilated and clean pen, are easier to feed and water and allow for much easier individual surveillance.
- Group pens allow for easier feeding, although no more than six calves per pen, making it easier to regularly observe the animals with one or two glances per pen.
- Healthy calves can tolerate quite cool conditions as long as they are protected from draughts and provided with a dry floor on which to lie.
- There are many types of flooring in calf pens and there is little difference in calf performance provided they can rest in a clean and dry location.
- Routine cleaning and sanitising of all feeding equipment is essential to maintain good calf health. The WATCH principle should be used when cleaning utensils and they should be allowed to dry completely before reuse.
- With the potential of many diseases being passed on from calves to humans, it is important to supervise children closely in the calf shed and ensure they wash their hands and faces before eating.

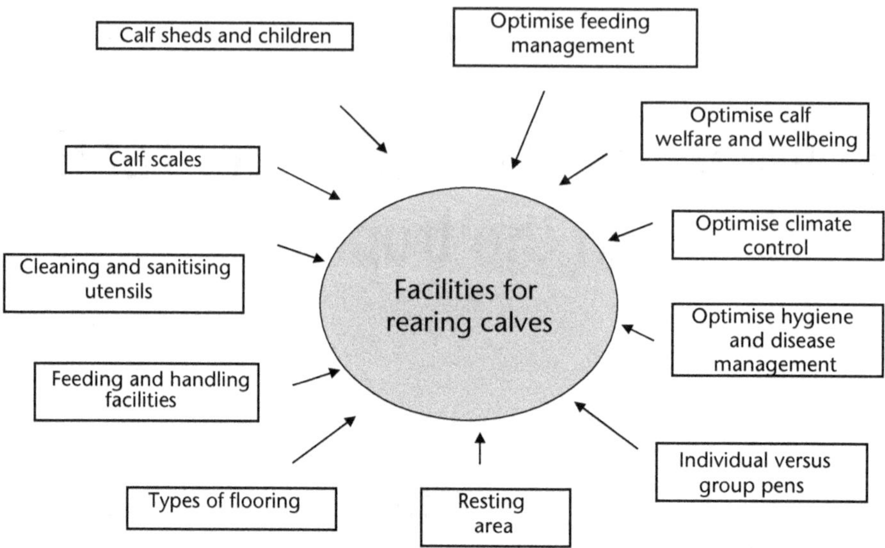

Figure 7.1. The key factors to consider when planning calf- and heifer-rearing facilities

This chapter discusses the housing of calves and heifers to ensure their wellbeing and acceptable performance during the milk-rearing and growing-out phase until they enter the milking herd.

Close attention should be given to the design, construction and maintenance of facilities for calves and heifers (see Figure 7.1) to optimise their performance in the herd with regard to:

- stock welfare and wellbeing
- feeding management
- climatic control
- hygiene and disease management.

7.1 Stock welfare

Throughout the world, public perceptions of farm animal welfare issues have the potential to markedly affect the sustainability of livestock industries, with national and international pressures likely to have increasing roles in determining how animals are managed. Because farm animal welfare is largely part of good animal and farm management, paying close attention to their day-to-day management should also ensure acceptable welfare.

7.1.1 The five basic freedoms of livestock

The welfare of cattle can be summarised in the 'five basic freedoms' as follows:

1. Freedom from hunger and thirst, through ready access to fresh water and a diet to maintain full health and vigour.

Calf crates allow for individual management and monitoring of newborn calves.

2. Freedom from discomfort, through provision of appropriate shelter and comfortable resting areas.
3. Freedom from pain, by prevention and, when sick, rapid diagnosis and treatment.
4. Freedom to express normal behaviour by providing adequate space, proper facilities and the company of other animals.
5. Freedom from fear and distress by ensuring conditions and treatment that avoid suffering.

All management and housing systems should be designed, constructed, maintained and managed to assist with these five freedoms.

7.1.2 The six basic elements of stock welfare

Specific to dairy herd management, calf and heifer welfare can be broken down to six basic elements: facilities and equipment; provision of feed and water; herd management; humane killing; staff competency; and preparation, selection, sale and transport of stock.

1. Facilities and equipment

These should be well designed, maintained and operated to ensure a high level of animal welfare, minimal stress or chance of injury. This includes:

- protection from extremes of weather
- provision of sufficient space with minimal possibility of injury

A well-constructed and managed calf-rearing facility.

- flooring designed to minimise slipping, falling and lameness
- facilities for water, feeding and restraint appropriately designed and maintained
- laneways, tracks and gateways designed to minimise stress and injuries
- equipment for euthanasia in good working order.

2. Provision of food and water

Food and water should be provided with consideration of environment, age and body condition that ensures stock health, wellbeing and productivity. That is:

- Stock should have access to food and water of an appropriate quality and amount.
- Stock should be fed rations formulated and balanced to provide the necessary nutrients for their desired level of performance.
- Short- and long-term plans should be in place to manage food and water shortages and drought.
- Practices should be in place to reduce the potential risk of toxicity or contamination during feed storage.

3. Herd management

Dairy calves and heifers should be routinely managed and handled to minimise stress, injuries and disease and promote good health and welfare. Such procedures should ensure that:

- Reproductive practices should be carried out competently to ensure good animal welfare and reproductive outcomes.
- Other routine husbandry procedures such as disbudding, castration and calving induction (if required), are carried out to minimise pain and unnecessary suffering.
- Systems to manage disease and other animal health disorders should optimise the planning, prevention and monitoring of the health of the dairy herd.
- Disease, injury, illness and stress should be identified and treated promptly with suitable expert advice sought as required.
- Animals should be inspected regularly to monitor their health and welfare.

4. Humane killing

Weak, ill or injured calves and heifers should be identified and treated appropriately or humanely destroyed using approved methods.

- Stock requiring humane destruction are identified and promptly euthanased.
- Competent staff are available to carry out humane euthanasia, using approved methods.

5. Staff competency

All staff responsible for managing and handling stock should be competent in their tasks and aware of their responsibility for good animal care.

- Staff carrying out routine husbandry, surgical procedures, reproductive procedures, administering health treatments and handling and transport of cattle are appropriately trained or experienced.
- All staff handling stock can identify signs of illness, abnormal behaviour or stress and ensure appropriate action is taken.
- There is a competent person either available on site or that can be contacted to handle emergencies and humane destruction when necessary.
- Staff behave in a manner to minimise fearfulness in cattle.

Stockmanship, plus the training and supervision necessary to achieve required standards, are the key factors in the handling and care of livestock. A management system may be acceptable in principle, but without competent, diligent stockmanship, the welfare of animals cannot be adequately safeguarded.

6. Preparation, selection, sale and transport of stock

Stock should be selected and appropriately prepared for transport to ensure they are fit for the intended journey.

- Preparation for sale and transport includes appropriate actions for feed and water curfews, identification and handling to minimise stress.
- Stock are selected for transport with consideration of age, class and condition to ensure they are fit for the intended journey.
- Any weak, ill or injured stock are not transported until deemed fit by a competent operator.
- Stock should be handled with care during loading and unloading.

7.1.3 Other general principles of stock welfare

Whether housed in cubicles, straw yards or cow sheds, in order to maximise performance and ensure satisfactory standards of welfare, the accommodation must provide for stocks' basic needs. As an absolute minimum, the housing must provide a comfortable, clean, well-drained and dry lying area together with shelter from adverse weather. It must allow the animal to move freely around without risk of injury.

Dairy stock at pasture choose to lie down for 12–14 hr each day, so a similar target should be achieved with stock in sheds. If they spend less time lying down, they are likely to spend more time standing in loafing or feeding areas, which can adversely affect hoof health.

Most farmers believe that animal welfare is just good cattle husbandry. The major welfare issues facing dairy farming include:

- housing and stock comfort
- castration of bull calves
- disbudding or dehorning
- branding
- transportation
- slaughter.

Lameness can be a major problem, both from the point of animal welfare and farmer profits. Farmers frequently recognise only 40–50% of lameness problems and these are often well advanced, making them difficult to treat. The calf or heifer's environment, both social and physical, and her ability to cope with it ultimately determine how bad lameness can become.

Early separation of calves from their dams is another welfare issue that is practised for animal health reasons, to decrease exposure to pathogens from the calving area. To improve the likelihood of calves receiving sufficient quality colostrum in their first 12 hr of life, they should be assisted to suckle their dam or the colostrum can be administered using a teat or stomach tube.

It is important to maintain animals in good body condition, house them in social groups in clean environments with adequate space to move around and rest comfortably, maintain feed, medication and production records and compassionately handle animals undergoing management practices, such as dehorning or castrating, and those that are injured or ill.

It should be remembered that, when working with calves in their first few days of life, they are newborn and have no understanding of what we want them to do. All handling should be done in a quiet and gentle manner and calves should not be carried by the legs, thrown, kicked, beaten, dragged along by their head or prodded with sharp instruments. At no time should dogs, sticks or electric prodders be used on calves of this age. If transported in a vehicle or trailer, they should not be overcrowded and be provided with at least 0.2 m^2/calf.

Most Western countries have strict codes of animal welfare covering transport, housing, handling and feeding. Farmers purchasing dairy stock from such countries may be expected to abide by the principles of that country's animal welfare codes. This is particularly the case with young stock.

7.1.4 Problems of confinement

Dairy stock imported from Western countries have almost invariably been reared under grazing conditions, hence have never been exposed to a continual shed environment, as is common on most small holder farms (Moran 2005). Compared with grazing, confinement creates specific problems such as:

- restricting opportunity to seek comfort; for example, if only provided with cement floors
- creating problems of high humidity, which can be more detrimental than high temperature
- limiting opportunity for exercise, hence the need for routine hoof trimming
- increasing exposure to infectious diseases
- other health issues, such as mastitis and uterine infections when hygiene is poor during milking and calving
- creating problems of heat detection for artificial insemination
- requiring greater efforts in sanitation
- magnifying problems of social dominance
- increasing capital investment.

7.2 Housing calves

The way calves are housed depends mainly on the climate. The colder the climate, the more attention should be given to housing, although calves are more sensitive to draughts than to the actual cold. Calves can be housed in single or group pens. Calf sheds need not be expensive structures, but they should be built in ways that allow for easy cleaning and maintenance of hygienic conditions.

Calf housing should be directed more towards protection against severe extremes rather than from normal seasonal variations in which the calves' hair coat and natural instinct provide protection. Good ventilation is essential to minimise potential problems with pneumonia.

7.2.1 Selecting the site and shed layout

A good calf house is one that meets the demands of both animals and operator at reasonable cost (Moran 2002). Calves obviously need protection from rain, draughts (in winter) and direct sunlight (particularly during summer), as well as a clean, dry floor on which to lie. The operator needs a building that is comfortable and convenient to work in, particularly for routine daily tasks such as feeding and cleaning. The operator should also be able to clearly observe all calves at all times (during day and night) and be able to single-handedly catch, restrain and administer treatment to any animal requiring attention.

Siting the shed

Consider the area chosen to rear the calves. The history of the area should be investigated and fully known. For example, if calves will spend considerable time outside the shed, that area could have been grazed by older stock and pose a high risk of disease for calves due to faecal contamination. Similarly, multipurpose sheds may have been used to store chemicals, other farm products or machinery, putting the calves at risk of poisoning or residue contamination.

If possible, a good plan is to maximise the time when the shed is empty between batches. All bedding should be removed and rails, gates and feeders should be cleaned to remove faecal material. Steam cleaners are very effective for this purpose. Disinfecting gates, pens and shed walls with a broad-spectrum disinfectant can also be beneficial. However, using products to sterilise dirt, such as lime, have minimal impact on pathogens in the soil and can be an irritant to staff and calves exposed to them; there seems little scientific justification for their use. The longer the calf-rearing area is free of calves and bedding, the less disease causing organisms will be present: even a 1 or 2 week period can significantly reduce the concentrations of disease organisms.

Sunshine provides a number of benefits to calves because it warms calves in winter and dries out and disinfects their bedding. It also stimulates vitamin D production (which is important in utilisation of milk). The shelter is considered adequate if it is comfortable for a person to sit dressed only in light clothing.

The building and fittings should be designed to allow for easy cleaning and disinfecting between batches of calves. Any sharp pipes, broken gates or sharp edges in calf pens should be eliminated to prevent calves from scratching and cutting their skin. The shed should provide adequate space for storage of feed, hospital pens for sick calves, a desk for record keeping and a washing-up area (with hot water) for milk-feeding utensils. Protection against vermin and birds, and possibly even flies, is important. The siting of the feed store and the location of doors and passages are important because these will influence the total distance covered each day. Therefore the shed should be designed to enable tasks to be carried out quickly and without unnecessary repetitive movements.

Ventilation

Good air quality is critical for good calf health. High levels of disease-causing organisms and pollutants such as ammonia and other noxious gases can lead to high levels of disease. High temperatures and humidity can increase the ability of disease-causing organisms to survive in the air.

Where cold stress can adversely affect calf wellbeing, the shed must be designed and positioned to be draught free at calf height, but still allow regular air movement above to remove pollutants. Improved ventilation can be achieved in sheds with their back to the prevailing weather by creating an opening of 60 cm at the very top of the wall. This is necessary in open-faced sheds that are 9 m or more deep. Sheds that are not purpose designed can pose some problems. Ideally, calves should have a near solid side on the prevailing weather aspect of the shed, allowing for ventilation gaps. Slatted floors can provide excessive draughts and should be modified with shade cloth and bedding to prevent this occurring in areas with cold winters. Many sheds have no windows or air vents and these should be modified to provide better ventilation and improved animal health and welfare outcomes.

A quarantine area for newly introduced calves could be incorporated, but this should by specifically designed for thorough cleaning and disinfecting, with sufficient pens to rest each one for at least 2 weeks before occupancy by a new calf. Pressure hoses should be used for cleaning, while a steam cleaner may reduce the need for disinfecting pens. There is less chance of disease build up by using galvanised iron, rather than wooden, pens and fittings because dried faeces are easier to remove from iron fittings. Poorly cleaned wooden pens can provide an ideal medium for bacteria to survive.

Rather than wean calves directly onto pasture, it is best to house them for several more weeks to ensure effective rumen development. Ideally, weaned calves should remain in their milk feeding pens for at least a few days after weaning so they can get used to their new feeding regime before having to adjust to any new pen and pen-mates. If the concentrate mix fed during milk rearing is different to that fed after weaning, both should be on offer (separately or mixed together) for the few days after weaning. If outside grazing areas are available, weaned calves and older heifers could be grazed, as long as they can be provided with troughs for supplementary feeding and shelter when required.

A good management practice is to divide replacement heifers into the following four age groups:

- >1 month-old milk-fed calves
- 1–5-month-old milk-fed and weaned calves
- 5–22-month-old weaned heifers and yearlings
- 22 months and older pregnant heifers.

7.2.2 Individual pens

It is easier to observe calves, and there is less risk of disease transfer, when they are individually penned. Ideally, the pens should be moveable and be enclosed with three draught-proof walls. Each pen should have 1–2 m^2 of floor space with a well-drained floor and, if desired, clean bedding each day. An individual pen 1.5 m × 0.8 m is the minimum required up to 4 weeks of age, while a pen 1.8 m × 1.0 m will suffice for an 8-week-old calf.

There should be containers for milk, water, concentrates and roughage. Calves should be confined to these pens until weaning, after which they could be allowed outside during the day. The walls should be impervious to prevent nose-to-nose contact and contamination by dung and urine. They should be constructed of material that allows thorough cleaning and disinfection between calves, such as sheet metal or plywood.

Alternatively, calves can be tethered in a 60 cm wide pen, which is open at the back. Dung and urine usually pass to the rear rather than to the walls of the pen, the next pen or the feed and water buckets. The walls can be constructed of cheaper material and less bedding is required. However, the bedding at the rear will require more frequent replacement and the pens should be raised 20 cm above the ground to keep them drier.

Some farms rear calves in individual hutches, which are usually placed outside, side by side, with each one enclosed on three sides. The design aims to provide ample ventilation without exposing calves to direct draughts. The main advantage of hutches is the improved natural ventilation and reduced chance of disease transmission from calf to calf. Even in very cold weather, calves in outdoor hutches eat more feed, grow faster and have less animal health problems that those inside warm barns. Furthermore, calf hutches can be moved to new areas with each batch of calves to reduce any disease build up. However, their role in calf rearing in the humid tropics would be limited by reduced access for feeding during lengthy periods of rain and the accumulation of mud when located on bare ground.

7.2.3 Calf cages

Containing each calf in a raised metal cage provides the best housing because they are isolated from each other, live in a well-ventilated and clean pen, are easier to feed and water (because the rearer does not have to bend down to feed them) and, of most importance, allow for much easier individual surveillance. Calves should live in cages for up to 3 weeks of age, after which they can be moved to individual pens.

Each cage should be 110 cm long, 75 cm wide and 105 cm high (including the legs). The rear of the cage is a removable gate to put the calf into, or remove it from, the cage. The floor can be made from wood or plastic, with the floor lattice perpendicular to the length of the cage. The space between floor boards should be 1.5–2 cm. The cage should stand on legs and be 30 cm above floor level. To avoid injuries, there should be no sharp edges inside the cage. The metal bars should be about 15 cm apart: any narrower can lead to injuries if calves get their legs stuck between the metal bars.

Three buckets should be attached to the front of each cage. For teat-fed calves, one is for water, one is for concentrates and one is for hay. For bucket-fed calves, the concentrate one should be cleaned out each day and used for milk feeding. If desired, straw can be placed on the cage floor for calf comfort.

Physical contact with other calves should be avoided, but calves need to see each other. Cages should be at least 50 cm apart and placed in a well-ventilated but not draughty place.

7.2.4 Group pens

It is more difficult to observe individual calves in group pens, but feeding them is easier. Slow-learning calves in group pens will often learn to drink from a bucket or nibble on concentrates more quickly than they would if in individual pens or cages. The floor should be well drained and can consist of wooden slats or deep litter where they might lie down. Troughs for water, milk, concentrate and roughages are required in each pen. Calves can be fed via teats or taught to drink from a bucket or trough. Self-sucking or sucking other calves' udders or navels can be a problem, but this is less likely with teat-fed calves.

When housing calves in groups, the optimum group size is six calves or fewer, with sufficient floor space (at least 1 m^2/calf), making it easier to regularly check the animals with one or two glances at each pen. Ideally, calves should not be moved from one group to another. Sick calves should be moved to hospital pens and, once recovered, to a new group. New calves should never be introduced into existing groups to replace dead calves.

With a 35 cm/calf pen frontage and 1.5 m^2/calf total area, group pens should be 3–4 m deep. In fully enclosed sheds, the stocking density is generally limited by the air volume per calf, rather than the floor space. The recommended cubic capacity is 6.5–7 m^3 per calf, with a ceiling height of at least 2.7 m.

7.2.5 Isolation pens

There should be some 'hospital pens' for isolating sick calves and heifers. This should be easy to clean and sterilise and have solid partitions to minimise contact between stock. It is convenient to store equipment in a specific veterinary drug box close by these pens for treating health problems, taking blood samples, and so on. There should be one treatment stall per 20 calf pens or cages.

7.3 Physical comfort of calves

The ideal temperature and humidity for calves is 17°C and 65% relative humidity. However, a normal healthy calf, eating well, is remarkably cold tolerant and is hardly affected by air temperatures below freezing point, provided it is dry and not exposed to draughts. On the other hand, sick calves, particularly emaciated ones with poor appetites, are very susceptible to cold. In regions with cold winters, the hospital pens should be in a warmer part of the calf shed and should be able to be heated, if required.

Calves lying on dry concrete lose more heat than those lying on wooden slats or damp straw. The warmest bedding is deep dry straw, wood chips or rice hulls. Draughts coming up through wooden slats or metal-grating floors should be eliminated during winter. These draughts can be easily detected using a lighted candle or match. On the other hand, draughts on hot summer days improve comfort by decreasing heat loads on shedded calves. Heat stress can also be reduced through constructing sheds with insulated roofs and well-ventilated walls, and by feeding calves in the cool of the evening.

Once air temperatures exceed 26°C, calves are outside their comfort zone. The degree of heat stress can be gauged by the calf's respiration rate. If it is more than 60 breaths/min, the calf is showing signs of stress. If it approaches 100 breaths/min, artificial cooling is required. This can be achieved by spraying with a garden hose or sprinkler system and fan, as with other stock in the herd. In hot, humid tropical regions, well-constructed cow sheds, with high roofs, open sides, sprinklers and fans will also be the best housing for replacement stock.

Plentiful supplies of fresh drinking water are essential for calves to cope with hot weather. Because stock can double their water intake once temperatures approach 30°C (compared with 15°C), water containers should be checked and refilled more frequently.

With regards cold stress, the lower comfort zone temperature is 13°C for calves from birth to 3 weeks of age and 1°C for dairy calves from 3 weeks old to weaning. Consider providing specific facilities for re-warming sick young calves, such as heat lamps or warm blankets, ensuring that attention is given to ventilation and sanitation following use of such a warming box. Calf coats can improve insulation of sick calves, while facilities to heat milk or drinking water can also provide additional methods to warm sick calves.

7.4 Types of flooring

The floor of calf pens is the surface on which animals stand, walk, lie down and pass excreta. It must – depending on the needs – be either solid, non-slippery and well drained, or comfortably soft, warm and dry and easy to clean by machinery. No single material meets all these specifications. Of most importance, it must provide calves with secure footing.

Wooden slats (50 mm × 25 mm with 20 mm gaps) placed 150–200 mm above a concrete sloping floor are ideal for calves. This arrangement allows the urine and dung to pass through the grating onto the concrete below where it can be easily removed by hosing without unduly wetting the calves. The grating should be made in sections small and light enough to be removed from the calf shed for thorough cleaning and disinfecting to prevent any build up of disease.

Wire mesh suitable for calf shed flooring is also available. Designed specifically for pig pen floors, it incorporates a mesh opening of 12.5 mm × 150 mm. It is welded onto a metal frame and should be positioned about 150–200 mm above the concrete floor.

During the milk-rearing phase, calves lie down for 17–19 hr every day. A deep litter of rice hulls or wood chips 40–50 cm thick over a concrete floor is probably the best bedding material. With occasional topping up and removal of excess dung, rice hulls can stay clean and dry for up to 4 months in summer. In winter, they will need replacing more frequently. Access could be provided for a tractor with a front-end loader or scraper to clean pens between batches. Tractor traffic must then be considered when planning concrete floor thickness. A damp-proof membrane should always be included in concrete floors. Sand does not make good bedding for calves because it does not provide any insulation and can accumulate in the stomach of calves that may consume it. The use of straw as a bedding should be avoided when such straw is also supplied as a feed source.

Alternative flooring could include concrete in the feeding area and a rice hull, wood chip or sawdust deep litter at the back of each pen. Dry straw is excellent, but is very labour intensive. When given a choice, calves seem to prefer sawdust and rice hulls to straw and wooden slats, and they least prefer metal gratings. Despite these preferences, there is little difference in performance of calves raised on either straw bedding or wooden slats.

Effluent disposal from pens is most important. A minimum fall of 1 in 20 will ensure that free liquid drains away. Drainage channels should run under the feed and water buckets at the pen fronts and drain both pens and passages.

7.5 Feeding and handling facilities

Feeding space requirements for individually fed calves (whether individually or group penned) is 35 cm/calf. This may limit the shape of group pens with bucket feeding to allow for sufficient frontage to the feeding passage. When feeding concentrates from a trough or bulk feed bin, allow 10 cm/calf. Troughs are more versatile, but less protected from birds and vermin. Hoppers must be robust and provide an even flow of feed without blocking or bridging. They should also be kept clear of the built up manure. A 10-week-old calf is able to reach into a trough 70 cm above the ground. If restricting or controlling concentrate intakes, 20–30 cm feeding space/calf is required. Group-fed weaned heifers also require 30–35 cm/head feeding space.

Large volumes of milk replacer can be mixed with warm water in stainless steel tanks using electrically powered rotors. In large calf-rearing sheds, it can be pumped or gravity fed to buckets using a petrol bowser dispenser. Feed scales are essential to ensure accurate weighing of powder. Whole milk can be pumped directly from the milking parlour to buckets or even to feeding drums located in nearby paddocks. If transition milk is being preserved, a milk line could be used to take it directly from the milking parlour to the preserving tank. Hot water is essential for cleaning feeding utensils.

Portable milk tanks with a delivery hose could be used when rearing calves in outdoor hutches. Whatever the method, regular flushing of hoses with water and thorough cleaning of milk dispensers and buckets is important. Concentrates can be automatically handled and dispensed using large silos and conveyors of the auger, chain or endless belt-type, such as those used for supplementing dairy cows during milking. Straw can be chopped then fed out in troughs or more easily handled unchopped using hay racks.

Water can be supplied through troughs or bowls, with one communal trough per two pens or one water bowl for up to 20 calves. Under normal conditions calves can drink up to 15 L/day, and 25 L/day on hot summer days. Drinkers must be guarded against pollution by faeces (both from calves and birds) and damage by rubbing. Water bowls and troughs should be of the low pressure type, because cattle tend to play with water. They should be sited away from the resting area so that spillage may drain away freely. If using buckets to individually feed the water, the water should be changed, and not just refilled, every day. In very cold weather, there can be benefits from feeding young calves with warm water. Ideally, different buckets should be used to feed milk and water; if using the same bucket, it should be washed thoroughly after milk feeding. It is worthwhile having extra buckets for water that can be rotated around the pens so all the water buckets can be scrubbed and thoroughly dried every week or so.

Nipple drinkers are a common feature in pig and poultry sheds and have been used successfully with young calves. The principle of these drinkers is that, when thirsty, calves simply push a metal nipple in the water line to extract the drinking water. They are cheaper to install and keep clean, but slow-learning calves may need to be taught how to operate them.

Stock-handling facilities should incorporate a calf race, head bail and calf scales, and lead to a loading ramp. A race for adult cattle can be modified for calves by adding a partition to make it narrower. This should be at least 1 m high and should reduce the width of the main race to 40–45 cm. In group pens, self-closing yokes at the feed face allow for easy restraining of calves for closer inspection or veterinary attention.

Passages between pens should be wide enough to allow buckets to be carried in both hands (1.2 m), be easy to clean and self-draining, have a non-slip surface, and allow easy access to the calf pens.

One very important item for any calf shed is a centrally located white board plus erasable marker pens. Managers can list jobs for staff, or details of any calves requiring particular attention. To assist in feeding and health management, every pen should be clearly numbered to remove ambiguity when recording feed intakes, calves requiring attention, and so on. There should be adequate lighting for any night-time activities and a lockable cupboard for veterinary medicine and even a small refrigerator for storing vaccines and other drugs.

7.5.1 Calf scales

Weighing scales are an essential component of good cattle-handling equipment. They are important for monitoring the growth of calves during rearing and ensure feeding management is sufficient for growing heifers to achieve target weights. Chest girths tapes and wither height sticks can also be used to assess calf and heifer development.

Cattle scales also allow dairy farmers to monitor changes in cows' weights throughout lactation. This will ensure that cows are being fed and managed properly to take advantage of their ability to utilise body reserves for milk in early lactation then replace it later in lactation. Changes in body condition score are a guide to this, but weight changes are the ultimate measurement. Scales can also be used to check the weight of bags of concentrate feed ingredients or bales of hay or silage, to assist in supplementary feeding programs.

7.5.2 Office and staff facilities

Maintaining good farm records is much easier in a farm office. An area in the calf shed could be dedicated to keeping records. It must have a desk and good lighting. It must be a quiet place in which to set up the office files (preferably in a filing cabinet) and computer and office supplies. The 'how and when' of keeping farm records depends on the person recording them. Computers are very convenient, but require money to purchase and skills to operate efficiently. Record keeping should be given as high a priority as other farming activities.

Suitable chairs and tables should be included for business meetings with service providers and other farm-related visitors. Farm staff should also be provided with space to eat and relax when off duty. This could include a shower and toilet, food preparation area, and storage for their work clothes.

7.6 Cleaning and sanitising feeding equipment

As SHD farmers are more frequently being penalised for poor-quality milk, they have become more aware of the principles of cleaning and sanitising of their milking equipment in the cow shed. But what about their milk feeding equipment in the calf shed? How often do farmers clean and sanitise it? Much of this equipment may be stored in conditions for ideal bacterial growth: namely moist, with no direct sunlight and with poor air exchange. Feeding milk out of dirty buckets and teats is a common way to spread scouring pathogens from one calf to another. Ideally, all feeding equipment should be cleaned and sterilised between feeds.

Cleaning removes residual milk from surfaces, while sanitising (or sterilising) removes bacteria from cleaned surfaces. The principles of good cleaning and sanitising can be summarised as WATCH, namely:

- Water: good water quality is important.
- Action: such as mechanical action with a scrubbing brush.
- Time: leave equipment long enough for the chemicals to work.

- **C**hemicals: match the chemicals for the job, detergents for cleaning and sanitisers for sterilising.
- **H**eat: chemical activity doubles every 10°C over 50°C.

The recommended cleaning and sanitising procedure for calf feeding equipment (Heinrichs 2002) should include the following:

1. Rinse equipment in lukewarm water (40–43°C) to remove leftover milk, manure and dirt. Organic material reduces the effectiveness of detergents. Do not use very hot water initially because this causes the milk proteins to bond to the equipment, especially plastics, instead of being washed away.
2. Wash equipment in hot water (60–82°C) using a chlorinated alkaline detergent. Soak the equipment for 5 min and then scrub all surfaces with a brush. Soap helps to loosen fat and dirt while chlorine kills bacteria. Hot water is essential to clean the equipment thoroughly. If the water temperature falls below 49°C, fat and soluble proteins come out of solution and stick to the equipment. Scrubbing loosens and removes minerals and organic soils from the equipment.
3. Rinse the equipment in warm to hot water (43–65°C) containing an acid sanitiser for 2–3 min. Hot water is needed to activate the acid, which rinses away soap and lowers pH on equipment surfaces for 12–14 hr. The low pH prevents bacterial growth.
4. Place equipment upside down on a rack to drain and dry. It should be dried completely before the next use to prevent bacterial growth. Using a rack allows air to circulate inside buckets or bottles.
5. If calf scours are a persistent problem or equipment remains wet between feedings, a pre-feeding sanitising step may be helpful. Allow the equipment to soak in sanitiser solution for 2–5 min before use. A possible sanitiser is 70 mL of chlorine bleach mixed with 3.8 L of water.

Other key points include:

- Always read and follow label directions.
- These procedures are also appropriate for cleaning animal health equipment, such as obstetric chains used with difficult calf births, drenching guns or hand milking equipment.
- Cracked or scratched plastic containers are very difficult to clean properly. Therefore replace buckets, bottles or oesophageal feeders if the scratching becomes excessive.
- Brushes designed for bottles or teats will make it easier to cleaning them properly.
- Clean feeding equipment immediately after use to prevent residues from drying.
- Isolate the milk feeding equipment used for sick calves and wash it separately to prevent it from contaminating other equipment.
- All individual buckets for milk feeding should be cleaned thoroughly after every feeding. Using individual buckets prevents calves sharing pathogens during feeding, but they cannot prevent bacteria growing on left over milk or saliva.

Some of the traps when cleaning and sanitising milk feeding equipment include:

- **Inadequate rinsing.** Detergents do not work efficiently in dirty wash water. If the wash water is too cloudy, then there should be better rinsing beforehand.
- **Using extra detergent following inadequate rinsing.** This does not work and can damage plastic equipment and human skin. It also increases the costs and adds to the chemicals contaminating the environment.
- **Wash water is not hot enough.** Detergents have to break through milk films, lift the milk components from the surfaces and break them into small enough particles to stay suspended in the wash water. This does not happen efficiently when the wash water falls below 50°C.
- **Water is too hard for detergents to work well.** If the detergents will not soap up, the fresh water may contain too many dissolved chemicals, so a water conditioner or softener may be needed. This can 'pay for itself' because less detergent will be needed.

7.6.1 Detergents and disinfectants

Detergents and disinfectants should be selected based on the job they must perform. Their effectiveness depends on their concentration, water temperature and contact time. Although increasing any of these also increases their performance, any deviations from the recommendations can reduce their effectiveness.

When cleaning both housing and feeding equipment, the important contaminants are organic matter and minerals. Detergents for cleaning feeding equipment are similar to those used to wash milking equipment by hand. Alkaline detergents remove both mineral and organic matter, although hard water may interfere with cleaning. A detergent compatible with the quality of water should be selected and the label's directions for concentration, temperature and contact time followed closely. Chlorinated alkaline detergents are most commonly used in developed dairy industries. Chlorine loosens protein, while alkali substances dissolve fat, protein and carbohydrates. Acid cleaners remove mineral deposits.

In general, disinfectants should be broad spectrum, non-irritating, non-toxic, non-corrosive and inexpensive. Other considerations include their effectiveness against specific pathogens, safety of use, residual activity and the surface to be disinfected (skin, metal, plastic, rubber, wood, etc.).

The effectiveness of disinfectants depends on several factors. First, the surface must be clean. Disinfectants can be inactivated by protein (or other organic material), the wrong pH and soaps. Next, they must be applied at the proper concentration and temperature and be allowed adequate contact time. The degree of contamination affects the concentration and contact time required.

As with all farm chemicals, precautions must be taken when handling cleaning and sanitising materials:

- Make sure that containers of chemicals have tight-fitting lids and spouts to prevent fumes and spills. Clearly label the containers, including the manufacturer's directions for use.
- Store the containers in a locked room to prevent children and unauthorised people for entering. The room should be cool and well lit.

- Use acid-resistant gloves, eye protection, face shields and protective footwear when mixing chemicals. Mix only in ventilated areas.
- Slowly add chemicals to cold water (never hot) and never add water to chemicals.
- Never mix chlorine compounds with other chemicals because this can produce deadly chlorine gas.
- Display details of the cleaning directions for each piece of equipment for temperature, amount of water, amount of chemical and contact time in clearly visible locations.
- Display emergency numbers (poison control and hospital) near telephones.
- Place an eye wash station near the chemical mixing area; if chemicals get into the eyes, flush with water for 15 min and see a doctor.
- Do likewise if chemicals touch the skin, and remove contaminated clothing.
- Rinse empty containers and dispose of them according to any local regulations.

7.6.2 Using household bleach as a sanitiser

Bleach is a weak solution of sodium hypochlorite and can be used as a sanitiser to sterilise clean milk handling equipment. When purchased for household use, the label often includes a 'shelf life' or length of time when it remains stable. It is important to store bleach away from light and in a cool place. However, it is only likely to be effective for several months.

The bactericidal effect (that is its ability to kill bacteria) of bleach is increased by:

- time: the longer the exposure, the better it kills
- concentration: the stronger the solution, the better it kills
- temperature: the hotter the solution, the better it kills.

So, its effectiveness in sterilising clean equipment is best when using a strong, hot solution and with as long an exposure as practical.

7.7 Calf sheds and children

Unfortunately, with ever-increasing stories about accidents and deaths on farms, many of which involve children, the owners and managers of calf-rearing operations must become more aware of the dangers for children in calf sheds. With increasing surveillance to comply with Occupational Health and Safety requirements, staff need to be protected from work-related accidents.

When purchasing chemicals for veterinary or cleaning/sanitising purposes (whether specifically for calf rearing or for other farm uses), ask for the associated material safety data sheets and store them in a secure, but easily accessible, place. These sheets will provide you and your doctor with information essential to treat any accidental spillages or swallowings, with the latter more likely by children. Keep all chemicals out of reach of children, and preferably in a locked cabinet. Place a first aid kit in the calf shed, in a nearby office or in the staff room and make all staff aware of its presence.

It goes without saying that all children, particularly those that live in towns or cities, love calves. When children visit the calf shed, keep an eye on them, either yourself or ask one of your staff to do so. Even small calves can become unsettled and injure a small child if they are unable to move out of the way.

Milk-fed calves, being monogastrics, carry many of the same diseases as humans. These diseases are called zoonoses. The most important ones include:

- *Salmonella*, *E. coli* and *Cryptosporidia*
- ringworm, mange and other skin diseases
- leptospirosis, which is more of a problem with adult cows
- Q fever.

In past years, doctors located in dairying regions of the US and Australia have noted a close association between outbreaks of specific pathogens causing calf scours and its occurrence in very young children. When children visit the calf shed, extra precautions should be taken with their personal hygiene. They should be made to wash their hands and face carefully prior to eating. Ideally, they should change their footwear and even clothes. If they frequently visit the calf shed, for example to assist with feeding the calves, they should have a pair of boots specifically for use in the calf shed. Elderly people, with a reduced immune system, could also be more susceptible to zoonoses.

8

Milk feeding of calves

This chapter discusses the diversity of decisions that have to be made when planning the milk feeding program for replacement heifer calves.

The main points in this chapter

- The nutrients from liquid feeds can be supplied though whole milk, colostrum (and transition milk) and/or calf milk replacer (CMR).
- If calves are to drink milk from a bucket or trough, they will have to be individually taught how to drink. Calves will instinctively suck from a teat.
- Automatic calf feeders are a new innovation in calf rearing, but are expensive and require a re-evaluation of pre-weaning calf management.
- The more milk fed, the less solid feeds consumed, the longer time to weaning and generally the more expensive the pre-weaning feeding program.
- Milk temperature, frequency of feeding, feeding mastitic milk and labour requirements are additional issues to be considered in planning milk feeding programs.
- Multiple suckling on nurse cows (continuous or restricted) is an alternative way to rear dairy replacement heifer calves on milk.
- Dairy farmers in Asia often successfully combine restricted suckling with hand or machine milking.

The diversity of climate, milk returns and concentrate costs throughout Asia have created a wide range of calf-rearing systems. The length of time, if at all, that calves remain housed, the method and level of milk feeding, the type of solid feeds offered and the age and weight at weaning therefore vary widely (see Figure 8.1).

The simplest rearing system involves putting young calves out to pasture, giving them access to trees or a simple shed for shelter and feeding them whole milk to appetite (*ad lib*) from a trough or feeding drum for up to 12 weeks of age, but with no additional concentrates. Such a system appears to work extremely well in temperate areas during warm, dry weather with calves grazing top-quality, spring pastures. In adverse weather conditions, or if pasture quality is sub-optimal, it can lead to older age at weaning and poor early post-weaning growth. Although labour and capital costs would be low, feed

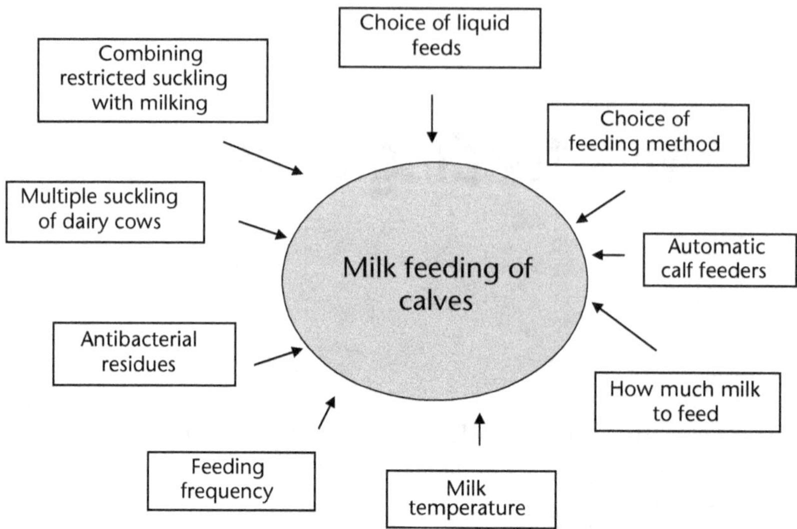

Figure 8.1. Decisions to make when considering milk feeding

costs are high, even if the whole milk has a low market value. Calf losses and disease costs may be acceptably low when the system operates effectively, but could be very high if it breaks down.

The other extreme would be to house calves for the first 2 months and feed them limited milk (or milk replacer), specially formulated concentrate mixes plus limited amounts of low-quality roughage. This encourages early rumen development and also achieves high pre-weaning growth rates. Following early weaning at 5–6 weeks of age, depending on concentrate intake, calves are still housed to allow greater control over nutrient intake. Once given access to forages, concentrates would be fed for several months to minimise stresses arising from the change in their basal diet to grazed pasture. Such a system maximises post-weaning growth rates and is the basis for achieving well-grown dairy heifers. Labour and capital costs are high, but feed costs are low or, at least, equivalent to the outside/pasture-only system. Calf losses and disease costs should be acceptably low, as long as the system operates effectively.

In both cases, it is the people who rear the calves, not the system (Moran 2002). The calf rearer teaches the animals to drink and decides on when and how the milk should be fed. This chapter describes many such systems, their advantages and disadvantages. It concentrates on practical issues, but these depend on the principles of calf growth and nutrition more fully described in previous chapters.

8.1 Teaching calves to drink

If calves are to be fed from buckets or troughs, they will have to be taught to drink. Because of their natural inclination, calves will instinctively learn to drink from teats, but will often need to be placed on the teat for the first few feeds.

A calf can be trained to drink from a bucket by backing it into a corner, standing astride its neck and placing two fingers, moistened with milk, into its mouth. As the calf

A slow way to introduce calves to milk.

starts to suck on the fingers, gently lower its mouth into the bucket of milk, taking care not to immerse the nostrils so it will not inhale the milk. Keep the palm of the hand away from its nose and as the calf starts to suck the milk, gently withdraw the fingers. Hold the bucket or have it supported about 30 cm from the ground.

This process should be repeated until the calf is drinking by itself or until it has drunk at least half a litre of milk. You may need to help the calf for several feeds. It is easier to train calves using warm milk, changing to cool milk only when they are drinking satisfactorily.

When training calves to drink from a teat, it should be attached to a tube that is filled with milk. As the calf starts to suck, lower the tube into the bucket of milk. The calf is usually able to keep up the supply by suction. It is easier for calves to learn to feed from self-closing teats, because milk remains in the tube between bouts of sucking.

8.2 The choice of liquid feeds

8.2.1 Colostrum and transition milk

The term colostrum is generally used to describe all milk not accepted by milk processors. However, transition milk is a more correct term for milk produced after the second milking post-calving. This milk no longer contains enough immunoglobulins to provide maximum immunity to calves, but still contains other components, which reduce its suitability for milk processing.

Milk from newly calved cows should not be put into the bulk milk vat for up to 8 days after calving. Regulations vary between countries and between different situations. During this period, a cow will produce considerably more colostrum or transition milk than that consumed fresh by her calf. If only rearing heifer replacements, the colostrum produced by cows that have given birth to bull and cull heifer calves, would then be available for milk feeding.

Using a 25% heifer replacement rate and 45 L of colostrum and transition milk per cow available for heifer rearing, this can provide up to 180 L of milk available per reared calf. These calculations take into account any milk used for early feeding of bull and cull heifer calves. There should be little need for dairy farmers to buy milk replacer or use marketable whole milk to rear their heifer replacements. Dairy farmers can save considerable money through modifying their transition milk storage systems to minimise the need to feed marketable milk or milk replacer to their heifer calves.

Transition milk has the greatest value when fed fresh or within a day or two of milking. It can be stored in a refrigerator for a week or so, or in a freezer for up to 12 months. In most farm situations, neither method is very practical for routine storage, except for a small supply of frozen colostrum for emergency use with newborn calves.

There is little difference in the immunoglobulin (Ig) levels in frozen compared with fresh colostrum. Only the first few litres of colostrum produced immediately after calving from older cows should be frozen for later use as a source of Ig. The ideal method to freeze the colostrum is in 1 L plastic bags placed in flat trays. This will produce wafers of colostrum about 2–3 cm thick, which can be rapidly thawed in lukewarm water. Very hot water should not be used to thaw the frozen colostrum because it can reduce its effectiveness in providing Ig.

Extremely bloody colostrum or colostrum from cows freshly treated for mastitis should not be stored, although it can be fed fresh to calves that are not to be sold.

Natural fermentation is an excellent way for storing transition milk for feeding as a source of cheap nutrients. It must be handled in clean containers to prevent contamination and should be kept in plastic or plastic-lined containers with lids. Old stainless steel milk vats are also ideal. If stored below 20°C, the natural fermentation will acidify the milk, stopping spoilage for up to 12 weeks. In warm conditions, preservatives may need to be added. These include propionic acid or formalin. The stored milk should be stirred every day to maintain uniform consistency and fresh milk should be cooled before adding. The preserved liquid will develop a characteristic odour, but calves will continue to drink it provided they are not abruptly switched from fresh milk or milk replacer to stored milk. They may refuse to drink it if it becomes too acidic. In this case, its palatability can be improved by neutralising it with sodium bicarbonate or baking soda at the rate of 10 g/L of milk.

Fresh colostrum has a slightly greater feed value than whole milk, so less can be fed or small quantities of warm water can be added to feed at the same rate as whole milk. When teaching calves to drink stored transition milk, it may be easier to begin feeding it warm – diluted with warm water (hot water will curdle it) – and then gradually change to cool, stored milk when calves are drinking more confidently. Calves will continue to drink such stored milk long after the rearer can't bear to get too close to it.

When the supply of stored transition milk begins to run out, fresh milk or milk replacer should gradually replace it over a week or so to give the calves time to accept their new diet. Changing from fresh milk or milk replacer back to stored transition milk can reduce intakes and lower growth rates.

8.2.2 Whole milk

Whole milk is the ideal food for calves. It has a high energy value and the correct balance of protein, minerals and vitamins for good calf growth and development. Health problems are generally lower when feeding whole milk compared with milk replacer because there is guaranteed quality control of the sources of protein and energy and there is no need to follow recipes to ensure the correct strength for proper feeding. Whole milk can either be the commercial milk being sold or it can be waste milk: that is, milk from treated cows or mastitic milk that cannot be sold.

Calves fed whole milk are less prone to scours than those fed milk replacer. Although it is common practice to feed mastitic whole milk to heifer replacement calves, recent evidence suggests that this could lead to an increased incidence in herd levels of mastitis in later years.

Whole milk and milk replacer can both be preserved by acidification for easier feeding management. Formalin can be added at the rate of 1–5 mL/L of milk or hydrogen peroxide at the rate of 5 mL/L of milk. Acidification can be achieved through adding 1.5 g citric acid/L of milk or including a buttermilk culture (or non-pasteurised yoghurt) to ferment the milk. If the milk is made too acid, calves daily intake will be reduced.

8.2.3 Milk replacer

To many producers, the decision on whether to feed whole milk or calf milk replacer (CMR) during rearing depends largely on cost. Sourcing a consistent quality of the milk replacer and its convenience for feeding are other factors influencing its use. Some farmers are concerned with the marked variation in milk replacer quality from batch to batch. Even though whole milk may be cheaper, it may not always be readily available for feeding to calves. For example, the calf feeding area may be some distance from the milking parlour. The composition of calf milk replacers and their feeding value relative to whole milk is discussed in Chapter 9.

8.3 The choice of feeding methods

To easily identify animals requiring extra assistance when drinking, young calves should be run together in small groups of no more than six calves. Some producers like to crate or tether their calves individually for the first few weeks to ensure all animals are drinking and for ease of observing signs of disease or poor performance. This also prevents the spread of disease between animals and, more importantly, between older and younger calves. This will also reduce the incidence of pizzle (or ear, navel and udder) sucking, which often occurs in very young calves run together in groups.

Running calves into individual stalls just for bucket feeding eliminates any problems of fast drinking calves poaching milk from other buckets. The use of self-closing yokes is an alternative method. Small or timid calves should be given the same opportunity to drink similar volumes of milk as bigger or more aggressive animals.

There are a variety of systems used for feeding whole milk or milk replacer. All will produce good calf growth and weaning weights if followed correctly. The major difference between any two systems is usually the result of the calf rearer rather than the system. Calves can drink from individual buckets or from communal troughs with or without rubber teats.

Buckets remove competition between calves for drinking space. By using one bucket per calf, each animal can receive a measured volume of milk, thus ensuring even milk intakes and calf growth rates. Small calves and timid drinkers can be given preferential treatment. However, labour requirements are higher and it is more time consuming than communal troughs.

To ensure the oesophageal groove will function properly and direct the milk to the abomasum, place the base of the bucket at least 30 cm above ground level where the calf is standing.

Troughs allow for feeding anywhere on the farm and not just in calf sheds. However, there is less control over individual milk intakes because calves drink at different rates and more aggressive calves have the advantage. Calves should be started on buckets then confined to a small yard to feed for a few days until they get used to trough feeding. Groups of calves will have more uniform growth rates when matched for drinking speed than for age or size. Each animal should be allocated a feeding space of 35 cm or, if using rubber teats, one teat per calf.

One innovative calf rearer in northern Victoria has modified 45 cm metal pipes into a series of troughs to ensure calves drink the same quantity of milk. He uses individual feeding stalls to allow only one calf per 35 cm space. Metal partitions have been welded into the pipe, limiting the volume of milk available to 4 L/calf (for once daily feeding). The trough rotates, so that when the milk is poured into it, it fills each compartment very quickly. He then rotates the trough upwards into the feeding position with a handle so the calves can drink their share of milk in whatever time it takes them. He has four feeders, allowing him to feed 80 calves in just 4 min. The troughs are easily cleaned with water and then rotated to empty and dry out.

Rubber teats give no additional nutritional benefit over bucket feeding because the speed of drinking milk has little effect on its utilisation. However, the production of saliva is greater in teat-fed calves and it may help maintain fluid intake in scouring calves. Teat feeding has also been shown to reduce the incidence of pizzle sucking in calves housed in groups. More capital is required in setting up the system and more labour is required for feeding and cleaning. Farmers often needlessly replace worn teats, but, as long as the teat can be kept clean, it does not matter if the end has been chewed off.

Farmers often prefer using teats into buckets because of the ease with which calves will learn to drink from teats. To many calf rearers, it seems illogical to provide both a teat and a bucket for each calf during milk rearing, because it doubles the cost of feeding equipment, greatly increases the time calves take to drink their allocated milk, then requires more labour to clean the equipment after use. Furthermore, it is easier for faster drinking calves to poach milk from their pen mates simply by pushing their mouth away from a teat than pushing their head out of a bucket. Calves can consume 4 L of milk from a bucket in less than 30 seconds compared with more than 1 or 2 min if using teats.

One way of feeding calves in groups using teats is with a suckle bar. This can be made from 50 mm PVC piping fitted with milk line entries and self-closing teats. Milk is poured into one end and sucked out by the calves. It saves carting milk and is easy to wash.

'Calfeterias' and feeding drums are used with rubber teats and can feed large numbers of calves quickly. Because the milk can always remain covered they can be fed away from shelter. The calf controls the amount of milk taken per feed so scouring is usually reduced as long as the total milk provided is consistent. They can then be used for *ad lib* feeding.

With calfeterias, the teats are either positioned in a metal frame, which is attached to the top of the milk reservoir with plastic tubes to draw milk from inside the reservoir, or the milk reservoir allows the milk to run into the teats by gravity. Modern calfeterias are made from moulded plastic to provide a reservoir of 2 or 4 L per teat. The teats in feeding drums are positioned around the top of the drum, while the plastic tubes nearly reach the bottom of the drum. The residual milk that cannot be sucked up into the tubes is usually left to ferment naturally.

Provided the milk is regularly stirred, the feeding drum only requires cleaning out once or twice each week. Even if the milk becomes excessively thick, cutting the ends off the teats will allow the calves to continue to suck up the milk. The milk must be stirred every day to ensure that it does not separate into a watery layer at the bottom with most of the protein and fat floating on the top. The tubes must also be regularly checked for blockages and build-up of milk deposits.

It is preferable to provide one teat per calf, although one teat for every two to three calves can be used with *ad lib* milk feeding. It is important to group calves on age and size to reduce competition if providing fewer teats than calves. Carefully watching of calves at feeding will soon identify the dominant animals and whether there are sufficient teats available. Poor 'doers' can be moved back to a lighter group of calves to improve their competitive ability. Groups should not exceed 20 calves (if using a 200 L drum) and the age range should be no more than 3 weeks.

Some calves, particularly younger ones, may lose interest and stop sucking before they get any milk. This problem can be overcome with self-closing teats or providing a pressure head of milk behind the teats; for example, the drum could be mounted on a stand and some teats positioned part way down the drum.

Suckling of milk directly from cows will be discussed later in this chapter.

8.3.1 Automatic calf feeders

In recent years, automatic calf feeding (ACF) machines have become popular on many large-scale calf-rearing operations. Calves can enter and leave the milk or concentrate feeding station at will, but their feeding regime is controlled by computer technology. Each ACF machine can handle four teats and/or concentrate dispensers, thus allowing up to 100 calves to be reared in a single group. They are promoted as labour-saving devices that can provide for a more carefully controlled milk feeding program. Each calf is individually identified to allow its milk feeding regime to be controlled by pre-determined programs of daily milk allocations. Some machines also allow for controlled concentrate feeding as well. The pros and cons of ACF are summarised in Table 8.1 (Moran 2006).

ACF technology is not cheap, because, as well as the initial capital investment of the machine and associated computer software, each calf will require an electronic ear tag and the calf-rearing shed has to be modified to hold larger groups of calves. Because calves can be reared in large groups, extra surveillance is also required to minimise issues with animal health and behaviour. The costs and benefits of ACF are summarised in Table 8.2. Their potential for integrating into calf-rearing systems in tropical Asia will be limited by the relatively low cost of farm labour and the need for extra management skills.

8.4 How much milk to feed

The major aim of calf rearing is to develop the rumen by manipulating intakes of liquid and dry feeds to the stage where calves can make efficient use of forages. The quantity of milk fed and the rearing system adopted should take this into account, while maintaining a balance between acceptable growth, cash cost and labour input.

As discussed in Chapter 3, the more milk fed to calves, the less solid feed and the slower the rate of rumen development. Because milk is a high-quality feed, the more milk drunk, the faster the growth rate. However, the efficiency of converting this milk to live weight declines as intakes increase. When fed *ad lib*, 6-week-old Friesian calves can drink up to 12 L/day, and Jerseys up to 9 L/day, of whole milk. By the time the calves reach 6 months of age, any live weight advantage in calves previously fed *ad lib* milk, compared with restricted milk, is lost.

With access to concentrates and good-quality forages together with once or twice daily feeding of 4–5 L of whole milk/day, Friesians should reach a suitable weaning weight (70 kg) in 9 weeks and Jerseys (60 kg) in 10–12 weeks. Many farmers still use live weight as their major criterion for weaning, often feeding more milk than is really necessary.

Although *ad lib* milk feeding is more expensive than other rearing systems, this system is often justified through faster growth rates and lower labour requirements, if using drum feeding. Earlier weaning compensates for the greater milk intake of *ad lib* fed calves and advocates of this system argue that it uses only slightly more milk over the whole period compared with restricted milk feeding. Provided that there are no setbacks to growth, weaning can occur at 6 weeks of age. Some farmers claim to be able to wean such young calves directly onto pasture, but it unlikely that rumen development would be sufficient and a severe growth check would be likely. If considering such a rearing system, calves would have to be fed 0.5–1 kg/day of concentrates at least until they are 10 weeks old.

Many experienced calf rearers in Australia initially feed milk twice daily at 10% the calves' live weight for the first few weeks and provide fresh concentrates and drinking water each day within the first week of age. This ensures each calf has a 'good start to life', with high immunity to diseases and a positive energy balance. Only then will the

Table 8.1. Pros and cons of automatic calf feeding (ACF) machines

Pros	Cons
Cost and labour issues	
Reduces labour input for milk feeding calves	Are expensive to purchase
Reduces area required for housing calves	Requires reliable supply of clean water
Rearer not faced with many hungry calves on entering shed each morning	Requires water-resistant power outlets (up to three per feeding unit)
Calves don't crowd around rearers when they enter yards	Requires good floor surface and drainage around feeding units
Less physical work involved in calf rearing	Labour still required for daily operation, cleaning and maintenance of ACF
Evens out work load during day	Need calibration with every new batch of CMR
Reduces human error in calf rearing (milk volume, temperature and concentration)	Need correct program settings
ACF machines generally feed calves more accurately than humans	
Rearing management	
Provides flexibility with milk feeding program, such as reducing intakes of milk or pellets at pre-determined stages	Calves are in larger than optimal group sizes
Can identify many (but not all) sick calves from reduced milk intakes	Requires greater surveillance of symptoms of poor calf health
Provides good performance data on each calf	Sick calves can more easily spread infections through contaminating teats
Can automatically weigh calves	Sick calves can more easily spread infections because of larger group sizes
Do calves really need warmed milk?	Must group calves on age or size
Allows easy tracking of calf performance, particularly important with employed labour	Must still routinely check calves several times each day
Can feed different levels of milk in same pen	Because calves are never fed to appetite at one feeding, rearer may become unduly concerned about calf appearance
Managing the weaning process	
Can computerise (hence standardise) weaning process based on pellet intakes	Some calves may be less inclined to eat pellets because they never really become hungry
Most calves more readily eat pellets than when teat fed (same when comparing manual bucket versus teat feeding)	Very young calves may not take to pellet dispenser easily, hence require pellet trough in pens as well
Calves may reach target pellet intakes for weaning earlier than in current system	Some calves may not reach target pellet intake for weaning until later than in current system
Individual calves can be weaned on their own performance	
Specific ACF issues	
Allows non-dairy farmers to enter calf rearing	Some computer skills required for correct usage
Once established, ACF technology makes it easier to rear more calves, hence potentially value add to a wasted resource (namely bobby calves)	Must routinely calibrate ACF for correct water and powder dispensing
	Powder measured in volume not weight, therefore could be inaccurate
	Some brands do not self clean, so extra time required for cleaning
	Need to train calves to take to teats without coaxing
	Rodents may live in ACF when not in use

Table 8.2. Associated costs and benefits of automatic calf feeding (ACF) machines

Costs	Benefits
Capital cost of basic ACF machine	Savings on labour costs
Capital cost of additional features	More controlled, easier weaning process
Durable flooring and drainage around each unit	Reduced shed space
All weather power supplies to processor, teat and concentrate feeders	Reduced purchases of manual milk feeding equipment
Increased cost of CMR and milk (if greater)	Reduced costs of CMR and milk (if less)
Higher animal health costs (if greater than with manual rearing)	Reductions in animal health costs (if lower)
Long-term adverse effects of poor calf performance (if poorer)	Long-term benefits of good calf performance (if better)
Cost of electronic ear tags (now compulsory with calf rearing)	
Annual maintenance costs of ACF equipment	

farmers restrict the milk or feed it once each day, which will stimulate concentrate intake and allow for a successful early weaning program.

As the sale value of whole milk or the purchase cost of milk replacer rises, there is increasing pressure for low-milk-feeding systems that still maintain good growth rates to achieve early weaning. This is possible by feeding only 4 L of milk/day, *ad lib* concentrates from the first week of age together with low-quality roughage. When calves are eating 750 g to 1.0 kg/day of concentrates, milk feeding can cease. This can occur between weeks 5 and 6. As with any system, milk feeding can be stopped abruptly or reduced steadily over the last week.

The quality and the palatability of the concentrate is the most important single factor in this system. It should be coarsely ground or pelleted. Inclusion of molasses or a sweetening agent can improve its palatability. To encourage early consumption, a handful of the concentrate should be placed in the bucket once the calf has consumed all the milk.

Ideally, calves should be individually penned until weaning because every calf should drink only 4 L of milk every day and increase concentrate intake to about 750 g to 1.0 kg/day before it can be weaned. If the calves are group fed such that the dominant calves consume more than their allocation of milk, they will appear more developed, but their rumen will be smaller and they will eat less concentrates than other smaller calves only drinking their milk allocation. Therefore group feed intakes are not a reliable indication of individual intakes.

After weaning, consumption of concentrates should increase to 2 kg/day until the animals are 3–4 months of age. Concentrates can then be gradually withdrawn, provided good pasture or forages are plentiful. This early weaning system is low cost and has minimal labour requirements once milk feeding ceases.

The total amount of whole milk fed during rearing can vary from more than 550 L of milk with *ad lib* feeding with no concentrates down to only 150 L plus 80 kg of concentrates with restricted once daily feeding. The implications of the threefold range in milk intakes at various milk prices will be discussed in Chapter 15.

8.4.1 Benefits of feeding large quantities of milk

Despite the more rapid gains achieved with feeding higher amounts of milk or milk replacer, there is little economic benefit from the more rapid growth rates and higher feed conversion efficiencies (Davis and Drackley 1998) in dairy heifers. In animals raised for meat production however, this can reduce time to market weight, hence improve economic returns. Such 'accelerated growth programs' also have a goal of earlier calving age (say at 18–20 months of age), but these programs rely on exceptional management.

8.4.2 Specific decisions for the tropics

Decisions on the most suitable milk-rearing systems for calves often depend on local circumstances. For example, in regions where ghee or butter is produced, skim milk and milk substitutes are becoming increasingly available in the tropics as livestock feed industries develop. What must be remembered is that the environment in most tropical countries is more hostile for the dairy calf than in temperate regions and that it is more economic to spend additional money on calf rearing in order to reduce calf mortality and increase live weight gains. It must also be remembered that most dairy calves in the tropics, whether from *Bos indicus* or *Bos taurus*, are smaller at birth and grow more slowly than they would in the temperate zones. Thus temperate feeding standards are not completely applicable in the tropics and, if adhered to, may lead to some overfeeding of calves.

Typical tropical dairy calves would weigh 20–27 kg at birth and during their first week should be fed no more than 2–3 L of milk/day. By the fourth week, this can be increased to 3–4 L of milk/day. Heavier calves should be fed correspondingly larger quantities. However, it is likely that the amount of milk should never exceed 4–5 L/day, because calves will begin to eat concentrates and forages in increasing amounts as they grow older. Calves should have continual access to good-quality drinking water.

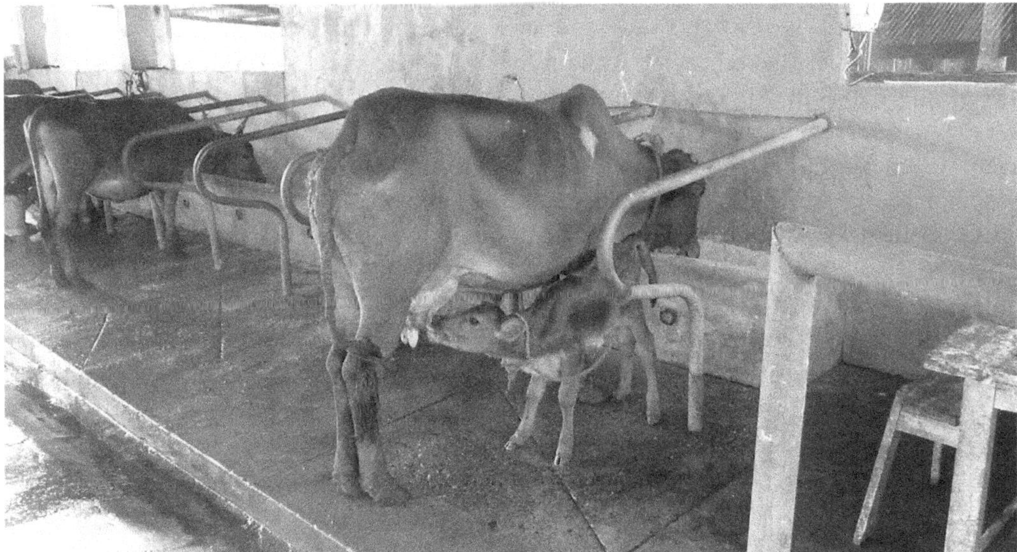

A calf suckling her dam prior to it being milked.

In the tropics, it is probably more economic to feed calves on additional milk than to attempt to early wean them at say 5–6 weeks of age. There are obviously many compromises with managing dairy calves in the tropics and 10 or even 12 weeks of milk feeding is more the norm.

8.5 Weaning age

The optimum age to wean calves off liquid feeds depends on many factors. These have been discussed above and in previous chapters. The most important factor is to ensure calves will continue to grow well once the primary site of feed digestion changes from the abomasum (for milk) to the rumen (for solid feeds). In other words, the calves have become fully functioning ruminants.

The age when milk is no longer fed should also depend on the quality of feeds available post-weaning. For example, Ibrahim (1988) suggested calves should be weaned at:

- 2 months, when the quantity and quality of roughage and concentrates are good
- 4 months, when the quantity and quality of roughage and concentrates are average
- 6 months, when the quantity and quality of roughage and concentrates are poor
- 8 months, when suckling and cows are dried off.

8.6 Other aspects of milk rearing

8.6.1 Milk temperature

The most natural way to feed calves is to teat-feed the milk at 39°C twice daily. Milk temperature is not important provided it is consistent from day to day. One exception to this would be feeding sick calves in winter, where they would have to use their own limited body reserves to reheat the milk in their digestive tract. It is certainly easier to train young calves to drink warm milk, after which it could be changed to cool milk. Very cold milk, removed directly from the milk vat, should be warmed, but only using small volumes of hot water prior to feeding. If the milk is too diluted, it may not clot in the abomasum, thus leading to digestive problems and possibly nutritional scours. If the milk is diluted, ensure that the calves are still offered the same quantity of whole milk.

8.6.2 Feeding frequency

Twice daily feeding is still the normal routine on many Asian dairy farms, but once daily feeding commencing after the first few weeks of life is adequate. Calves grow equally fast on either frequency when fed the same level of milk each day. Because the competition for milk may be stronger with once daily milk feeding, it is essential that each calf gets its fair share of milk. Correct grouping of calves is very important if feeding from a communal trough, as is at least one teat per calf if using rubber teats.

It is important to provide access to concentrates within the first week and to ensure it is fresh each day. Clean water must also be on offer because calves will drink more water than when fed twice daily. It is possible to strengthen milk replacer mixtures (that is, use less water in their formulation) to ensure smaller calves can still consume enough nutrients when fed the larger volumes once daily.

Calves fed only once each day will eat more concentrates at an early age because they have more time to get hungry and seek out other feed. Furthermore, calves can be fed at the most convenient time of the day, rather than after morning and afternoon milking, as is necessary when feeding twice daily. Once daily feeding should not reduce the frequency with which calves are inspected.

8.6.3 Milk dilution

Farmers sometimes dilute milk, either to warm it or as part of a treatment for scours. Dilution of milk or milk replacer reduces the intake of nutrients due to the calves' limited gut capacity. Some farmers even dilute milk when weaning calves so the animals will have the same volume but less milk solids. Calves can, in fact, be abruptly weaned off milk with no serious after effects.

The clotting of whole milk in the abomasum can be reduced in calves fed very diluted milk. Apart from warming very cold milk prior to feeding sick calves, there seems to be little benefit for calf feeding in diluting whole milk or reducing the concentration of CMR below that recommended on the bag.

8.6.4 Antibacterial residues

It is essential that calves sold for slaughter do not contain any antibacterial (or antibiotic) residues. Baby calves destined for slaughter at a week of age should not be fed milk from cows treated with antibiotics unless the required withholding period for each chemical is strictly observed.

The withholding period is the time following treatment, during which products derived from any treated animal should not be used in food production. This varies for particular drugs, with the route of administration into the cow (injection, oral or intramammary) and the dose rate. For most antibacterials, the withholding period for sale of milk is considerably shorter than that for sale of meat, which can be up to 30 days from administration. To be on the safe side, consider 30 days as the minimum withholding period for calves fed milk from cows given intramammary drug treatment.

Antibacterial compounds get into calves from four main sources:

- sick calves that have been treated, usually for scours
- healthy calves that have been fed using equipment contaminated with antibiotics
- calves suckling cows that have been treated with intramammary preparations or by injection
- calves consuming antibiotics through suckling cows that still contain 'dry cow therapy' preparations at calving, usually due to failure to massage the preparations into the udder when initially administered, or if the cow has only had a short dry period.

Calves that are intended to be reared as replacement heifers, but fail to thrive, are often sold along with other bobby calves. These calves are a particularly high risk group for antibacterial residues because they will often have been treated for some illness.

Calves are often sent to slaughter within days of being treated for scours with antibiotics or sulphonamides. In many cases, treatment with an antibacterial drug may not be necessary. Electrolytes, glucose and fluid replacement are the important components of an effective treatment for scours in calves. Antibiotics and sulphonamides should only be used on the advice of veterinarians, and withholding periods are as long as 28 days for some sulphonamide calf scour tablets.

8.6.5 Feeding mastitic milk

Milk from cows after antibiotic treatment for mastitis or other bacterial diseases cannot be sold and must be discarded. Estimates in the US are that this can amount on average from 20 to 60 L/cow. With the high incidence of mastitis in tropical small holder systems, this volume would be even higher in Asia. Feeding this milk to calves is one way to capture some economic value from an otherwise wasted resource. This milk is sometimes called 'blue milk', because of the blue dye used in intramammary mastitis treatment to colour the milk as an additional reminder that it is contaminated and should not be mixed with the market milk.

Controversy still exists as to whether feeding this milk to replacement heifer calves increases their likelihood to mastitis in later life. If calves are individually penned, there is no evidence of increased mastitis. The antibiotic does not adversely affect milk digestion, increase the likelihood of greater antibiotic resistance in future disease outbreaks nor have any long-term detrimental effects on production or health. There is conflicting evidence on the potential of mastitis bacteria to increase the incidence of future mastitis in group-fed calves that can suck the developing mammary glands of other heifer calves. For this reason, farmers may wish to discard it or feed it to male calves.

Mastitic milk should not be fed to sale calves without due regard to the withholding period of the antibiotic. Calves should not be fed milk from cows with mastitis caused by *E. coli* or *Pasteurella* unless it has been pasteurised.

Concerns about viable pathogens in waste milk have led to large dairies in the US installing pasteurisation plants to treat all whole milk to be fed to calves. An assessment of the costs and returns indicates that such plants would need to be used to feed 300–400 calves before becoming economically feasible.

8.6.6 Labour

One of the major factors influencing the choice of feeding method is its labour requirement. The time taken in milk feeding and washing can vary from half to 3 min per calf per day and even longer in inefficiently run systems (Moran 2002). One of the quickest systems involves *ad lib* feeding of naturally fermented whole milk from a series of feeding drums or troughs for large numbers of calves run together in a paddock or pen. In contrast, twice daily bucket feeding of milk replacer for small groups of calves in a shed is one of the slowest.

If feeding time can be reduced by 1 min per calf on a farm rearing 60 calves, that amounts to 1 hour less labour each day of rearing. Remember that reducing milk feeding to 5–6 weeks rather than the more usual 8–10 weeks also considerably reduces the total rearing time per calf.

8.6.7 Pizzle sucking

The problem of pizzle sucking can be common among artificially reared calves. Young calves are instinctively curious and, as well as drinking, eating and ruminating, they use their mouths for all sorts of apparently abnormal behaviour such as licking and chewing inedible (or unswallowable) objects, sucking the ears, navels, teats, tails and pizzles of neighbours, and even drinking urine.

Cross-sucking is a potentially dangerous way of spreading infection, while urine drinkers tend to show abnormalities of rumen development. In intensive rearing systems such as group-fed veal production, these calves invariably show slow growth and poor feed efficiency.

The incidence of calves sucking each other can be reduced by providing greater opportunity for them to satisfy this desire, such as feeding with rubber teats rather than buckets and using *ad lib* feeding drums or automatic milk feeders to give calves continual access to milk.

Hanging a piece of chain in the pen of group-housed calves may also be effective. Pizzle-sucking calves can be individually penned or tethered during milk feeding then offered concentrates immediately they have finished their milk allocation.

8.6.8 Trying out a new system

Whenever farmers visit other farms, they generally look to see how 'things are done' and may consider changing their practices to include any potential improvements they have seen. This is fine as long as they can be confident it will improve productivity and profitability on their farm, or maybe even 'make life easier'.

When considering changing some aspect of calf rearing, rearers have an ideal opportunity to closely compare the 'old' with the 'new'. They should be encouraged to change practices in just one or two pens and see how the calves perform in comparison with their existing system. But they must make sure they are comparing 'apples with apples'. For example, if changing to an early weaning system, it would seem logical not to compare calves at different weaning ages, but at the same age, when their rumens are fully functioning. Using live weight at 12 weeks, or even older, is the best way to compare different milk feeding practices.

8.7 Multiple suckling using dairy cows

The cheapest way of feeding whole milk to calves is to allow them to harvest it themselves by suckling cows. The ratio of suckler or nurse cows to calves should be adjusted so that each calf receives at least 4 L/day of milk. The milk production of the nurse cows should then be checked to ensure adequate milk supply for her calves.

Once one batch of calves is weaned off the cows, another batch can be multiple suckled. Growth rates of suckled calves are as good as, or even better than, those achieved with artificial rearing but they can be more variable because there is less control over individual calf intakes. There is a serious risk of infecting calves with certain diseases carried by cows. Multiple suckling should not be used for rearing heifer replacements on dairy farms where Johne's disease has been identified or where there is a high threat of the disease. *Coccidia* and *Salmonella* organisms can also be transferred to calves through close contact with mature cows.

Obviously, cows with active mastitis infections should not be suckled because calves can transfer the mastitis-causing organisms to other teat quarters and also to other cows. However, mastitic cows destined for sale could be used to foster bull calves for meat, provided the infected cows are isolated from other cows.

Nurse cows and heifers generally produce more milk, while heifers can reach peak milk yields quicker than if hand or machine milked. Research has shown that cows foster rearing two or three calves for the first 8–12 weeks of lactation often produce more milk when returned to the milking herd than cows run in the herd from calving. Nurse cows also seem easier to break into the shed routine after a short period of suckling.

There are two types of systems for multiple suckling: continuous or foster suckling and restricted or race suckling.

8.7.1 Continuous suckling

This involves fostering extra calves with the cow's own calf. All calves should be matched for age, size and vigour. A proportion of cows will not adopt other calves, and such calves will steal milk from more cooperative cows. This will reduce their milk supply to their own foster calves and can lead to variable growth rates in both groups of calves. In one instance, nurse cows rejected foster calves and, probably due to an increase in milk supply, her own calf died from scours.

For continuous suckling to work, each cow and her adopted calves must become bonded as a family unit so that the nurse cow will accept all her own foster calves, but still reject others. Once this bonding has been established, it is difficult to introduce a new calf into the family, for instance, to replace one that had died.

To help develop this bonding, the cow and calves should be kept together in a small paddock for about 10 days and the cow should be restrained in a race or bail daily for 3 days at feeding to make sure all her calves have been accepted. It is sufficient to starve the foster calves for about 24 hr and then constrain the cow, unmilked for about 12 hr, with the calves for an hour each time. Other mothering systems involve keeping the calves in small pens, then locking the cow in with her calves for an hour or so every day for the first week.

It can help if the nurse cow becomes confused after calving about which is her own calf. It can be removed and replaced with other calves that have been smeared with a strong smelling substance, such as neatsfoot oil, which has been placed on the cow's muzzle and also on her own calf. Some farmers use rope or a swivel chain and collars to tether one or two foster calves to the cow's own calf for a few days; in this case, the calves should be no more than 30 cm apart.

To maximise growth and rumen development, the calves can be given access to quality feed (grass and/or supplement) by creep grazing using electric fences. When weaning some calves early, they must be the adopted calves because the nurse cow could reject them if her own calf was removed first.

Foster suckling has the advantage that the cows can be run away from the dairy, leaving closer paddocks for the milking herd. It also allows the continued use of good breeding cows past their prime as milkers, low-testing cows or cows that do not fit the daily routine (for example, because of temperament or milking speed). Such cows have been known to milk for 18 months and rear a dozen or more calves. However, their calves tend to become wild because of lack of regular human contact and they may be difficult to train for milking. Nurse cows are less likely to cycle and this increases the spread of calving in seasonal calving herds.

8.7.2 Restricted suckling

The second system involves separating the calves from the cows, except at milking time when they are brought together in a small yard or in a suckle race. Up to four calves can suckle any one constrained nurse cow. A suckle race will restrict movement of cows better than a yard and hence allow smaller calves better access to available teats. The race can be made of 50 mm galvanised pipe construction, 75 cm wide, with one rail each side 76 cm off the ground. Moveable pipe barriers can be inserted into the race every 1.8 m to separate the cows. The floor should be concreted for at least 1 m outside both sides of the race to prevent the ground from becoming boggy.

To minimise teat damage, suckling should be limited to 15–20 min per session and cows should only be suckled for 3–4 weeks at a time. All quarters of each nurse cow should be suckled dry. Scours can be more of a problem with suckled calves because of the increased likelihood of overfeeding. Hygiene problems are all eliminated because the milk is harvested directly from the cow. It is important to group calves on age and size to reduce competition. With very high-yielding cows and large numbers of calves to rear, it is possible to divide the calves into two groups and feed each group only once each day.

There may be little saving in labour compared with artificial rearing because calves have to be brought from the paddock or calf shed to the milking parlour each time. Cows have to be selected – such as mastitic and freshly calved cows – and then drafted from the rest of the herd. Some cows are difficult to train to accept calves, such as those that continually kick. Others are better suited for restricted suckle rearing than for machine milking, such as cows with three functional teats, poor udders or slow milkers. Cows can also be rotated between the dairy and the suckling race and still run together in the milking herd. Because nurse cows produce more milk, they could lose more weight in early lactation and hence may require better feeding than those being machine milked.

Variations to this system are to allow the cow's own calf to suck her dry after each machine milking for the first week after calving. Alternatively cows and heifers can be race suckled each afternoon by fewer calves and then machine milked each morning. These variations prevent milk accumulation in the cow's udder, which can have a detrimental effect on yields later in lactation, while rearing several calves. The improvement in milk yield after these calves are weaned generally compensates for the milk previously taken by the calves.

Nurse cows do not begin to show oestrus after calving as soon as cows that are machine milked. To maintain a 12-month calving interval, calves should be removed from the cow for 24 hr about 8 weeks after calving. Cows will normally show signs of oestrus within the next 7 days and can be mated at this or the next oestrus 21 days later.

Early weaning requires strict rationing of milk, so it may be difficult to combine this with multiple suckling. However, this can be done successfully by 5 weeks of age by gradually reducing either the time of access to the cows or the number of nurse cows. Calves should be weaned onto good-quality pasture together with 1–2 kg/day of concentrates. The protein content of the available pasture should determine whether the concentrate is boosted with additional protein or is basically an energy supplement.

8.7.3 Combining restricted suckling and hand or machine milking

This is a popular way to rear calves in many tropical countries, particularly those in Central and South America. It is common for farmers to use dual-purpose cows and bulls to provide farmers with a good milk income as well as rearing a beef type calf with a good market value. On small holder farms in Thailand and Africa, calves milk reared with restricted suckling grew faster, used their milk more efficiently and had fewer health problems than did those bucket reared. Furthermore, the suckled cows produced more milk and had less mastitis than cows milked out by hand or machines.

However, this practice can be associated with an increased number of days between calving and first oestrus. It is often considered that milking cows with a relatively high degree of Zebu breeding require the presence of a suckling calf to initiate milk let-down. Therefore, there seem to be many benefits of rearing calves using restricted suckling. Indeed, this is often the norm on South American dairy farms, where dual-purpose cows are milked.

However, there are high labour requirements to match up cows with their calves prior to milking and the commercial value of whole milk is generally higher than for CMR. This means that suckle rearing is less practical on most Asian SHD farms, where many of the milking cows are grade Friesians that can easily let down their milk without the presence of a suckling calf. Even with purebred Zebu milking cows, it is possible to machine or hand milk them in the absence of a calf, through training the cows and culling those cows (say at the end of their first lactation), which will not let down milk without a calf present. Concern about poor milk let-down in the absence of any suckling calf has led farmers to use oxytocin injections routinely at every milking.

8.7.4 Pros and cons of restricted suckling versus artificial rearing

Pros:

- more milk is extracted with suckling
- the milk has a higher milk fat content
- reduced mastitis
- calf behaviour is considered to be more 'normal'
- will this milk be otherwise extracted and sold?

Cons:

- delayed oestrus unless better feeding management (due to live weight loss/reduced body condition or entirely nutritional)?
- increase in daily labour and management
- because calves are generally grouped, rather than individually reared, they require increased disease surveillance.

9

Calf milk replacers

This chapter discusses the considerations to make when deciding on a calf milk replacer (CMR) feeding program.

The main points in this chapter

- Despite their convenience when milk feeding calves, very few tropical dairy farmers use CMR as an alternative to feeding whole milk.
- Very cheap CMRs are all too frequently of poor quality.
- CMR must be made from quality ingredients and various visual criteria can be used to assess its overall quality.
- The fat and protein content of CMRs can be used to quantify their nutritive value relative to whole milk.
- This should be used to decide on cost relative to the value of whole milk.
- It is important that farmers understand the mixing strengths when preparing CMR for feeding calves.
- Milk-fed calves require about 500 g/day of milk solids.
- CMR and whole milk behave differently during digestion in the abomasum, so require different feeding protocols.
- If fed too frequently, CMR can lead to abomasal-induced milk bloat.
- CMR can also be used to boost the concentration of whole milk.

Calf milk replacers (CMRs) provide a convenient way to feed pre-ruminant calves. They can be stored long term as powder and mixed with water just prior to feeding. Calves can then be milk reared anywhere and at any time without having to source liquid whole milk. Provided the CMR is formulated correctly from good-quality ingredients and fed according to the instructions, which are usually on the CMR bag, calves can grow equally well when reared on CMR and their rumens can develop just as well as they would on a diet of whole milk.

Because manufacturing CMR directly from whole milk is an expensive process, and because whole milk has a high market value, the bulk of the ingredients for commercial CMR are either by-products of dairy processing or non-dairy products. Obviously, the

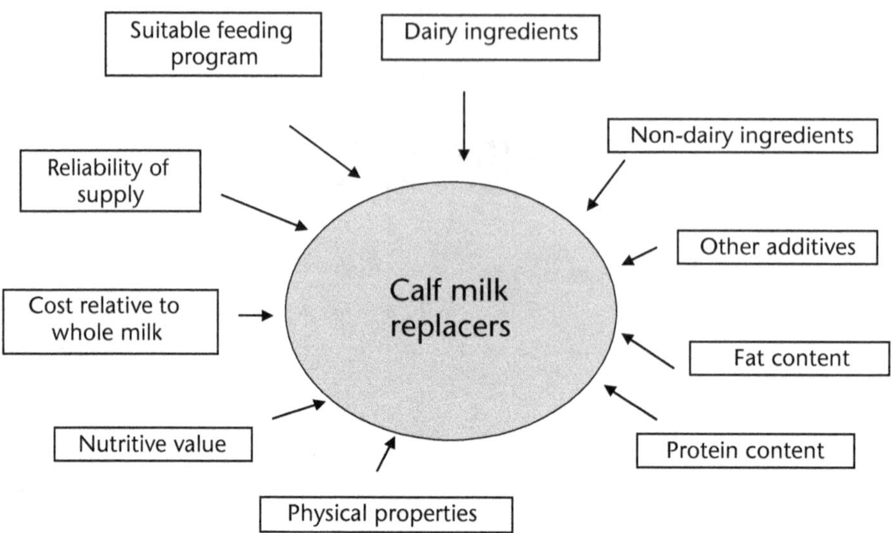

Figure 9.1. Considerations when selecting a suitable calf milk replacer

cost of CMR must be competitive with whole milk for dairy farmers to consider using it when they have copious supplies of fresh milk from their own cows. Despite this, most dairy farmers in the tropics only use whole milk in their calf-rearing systems. Many of them are unaware that CMR can be a cheaper form of liquid feed. Others may have tried it but had a bad experience (say from a batch of poor-quality powder or a limited supply), so then choose only to use whole milk in their rearing systems. Because they do not have to pay cash for their whole milk, in contrast to purchasing CMR, dairy farmers often perceive CMR as an expensive alternative.

In many tropical countries, the supplies of CMR are often unreliable, while its quality can be very variable. Much of the CMR fed on these small farms is imported, mainly from Europe, so it is likely that this should be a quality product, because the CMR manufacturers would have to produce a quality product for their domestic market. Some of the economic factors involved in feeding CMR compared with whole milk are discussed in Chapter 15. Considerations when selecting a suitable calf milk replacer are summarised in Figure 9.1, while a detailed manual on preparing and feeding of milk replacers (and starter feeds) has recently been published by Krishnamoorthy and Moran (2011).

9.1 The composition of milk replacers

A good-quality milk replacer should be similar in chemical composition to whole milk. It should contain the nutrients that calves can digest and in the right proportions. Most milk replacers form a clot in the abomasum and so provide a slow release of nutrients to the duodenum. There are others that do not clot in the abomasum and are primarily digested in the intestines.

Milk replacers are generally formulated from by-products of dairy processing, together with animal fats plus added vitamins and minerals. Whole milk powder consists

Farmers learning how to mix calf milk replacer solution (Vietnam).

mainly of lactose (36–40% of DM), fat (30–40% of DM) and milk protein (28–32% of DM). The protein is principally made up of casein, but also includes the whey proteins, albumin and globulin.

The by-product of butter making is skim milk, which consists mainly of lactose and all the milk proteins; it has only half the energy value of whole milk. Whey, the by-product of cheese making, consists only of lactose, albumin and globulin, and is even lower in nutritive value. When used as the basis of milk replacers, additional fats are required.

Commercial milk replacers usually contain 20–24% protein. Young calves can only digest proteins of milk origin, such as those from skim milk and buttermilk powders. The degree of processing of these powders affects the calves' ability to digest this protein. Excessive heating denatures the protein, leading to poor clotting in the abomasum and rapid passage of milk into the duodenum. Spray-dried milk powders, manufactured at lower temperatures than roller-dried milk powders, are the preferred source of powder for milk replacers.

Milk replacers should contain 15–20% fat and the type of added fat used will influence its utilisation by the calves. Tallow (a by-product of abattoirs) is the most common fat to include because vegetable oils, which contain high levels of polyunsaturated fats, can cause scouring in young calves. Tallow is preferred because it has a similar fatty acid composition to milk fat and is cheap. Tallow is one of the few animal by-products that can be fed to ruminants. The fat must be incorporated carefully so that the powder dissolves easily in water and the fat globules become sufficiently small

A calf waiting for its feed of calf milk replacer.

so that they do not separate out in the solution following mixing. Lecithin is usually included to assist with the incorporation of added fats and to improve their utilisation in milk replacer powders.

High-quality milk replacers have a fibre content of less than 0.1%. Fibre originates from plant material commonly used to increase protein levels in milk replacers. For every 0.1% increase in fibre content in replacers, about 10% of the total protein has been derived from plant, rather than milk, sources.

A typical milk replacer contains 70–80% milk solids, 17–20% animal and vegetable fats (for example tallow), 2% lecithin, traces of minerals (copper, zinc, manganese, cobalt, iron and iodine) and vitamins (A, D, B_{12}, K and E) with added antibiotics or antibacterial drugs.

The inclusion of antibiotics in milk replacers is a matter of concern, particularly to producers rearing their own calves born on-farm. New diseases, such as a different type of scour-causing bacteria, can be introduced through bought-in animals. This is why antibiotics are added to some replacer powders. In theory, calves should not routinely be given antibiotics because the sooner any disease outbreak can be identified and diagnosed, the sooner the calves can be treated. Low-level antibiotic feeding will mask a low level of disease, so that by the time calves show any symptoms, more intense treatment may be required. Furthermore, regular use of antibiotics will increase the risk of cull calves being sold for slaughter with detectable levels of antibiotic residues in their carcasses.

Most cases of scours are caused by poor feeding management, rather than infectious agents, so antibiotics, which will not be effective against viral or protozoal scours anyway, serve little purpose in most cases. By continually feeding antibiotics to calves, bacteria can develop resistance to them. This means that, if a bacterial disease does break out, the antibiotics prescribed by the veterinarian may not be able to control the resistant bacteria.

Antibiotics are also added to milk replacers to stimulate feed intake. Because antibiotics deliver the greatest improvement when management and hygiene are not the best, their routine use can give a false sense of security, which is followed by a generally poor job in calf raising.

Powders based on milk by-products are expensive and attempts to reduce their costs through using alternative protein and energy sources have been largely unsuccessful. Soybean or soya flour is a vegetable protein by-product successfully fed to older animals, but it contains an antigen that inhibits protein (in this case, trypsin) digestion in milk-fed calves. This anti-trypsin antigen can be destroyed by heat treatment prior to inclusion in replacer powders, but results from calf production trials to date are not promising.

Calves cannot digest starch in their diet until their rumen is functioning. As little as 2% starch in milk substitute diets will depress growth and increase scouring in very young calves. CMRs with high levels of starch are not suitable for such animals. The content of starch and the proportion of milk protein to total protein should be detailed on the CMR bag.

Long-term storage of CMR powders is important. They must be packaged properly to keep out air and moisture. They should be vacuum sealed in a plastic bag then enclosed in a light-proof bag. Even with this protection, they are best used within 6 months of purchase. Good-quality powders include an antioxidant to reduce the deterioration of fat during storage.

9.2 Describing quality in milk replacers

In the 1970s, a panel of Australian dairy specialists developed a set of standards for milk replacers to ensure their suitability for calves less than 3 weeks old. This has been updated by Heinrichs (2002). These standards were as follows:

- The powder should contain between 15% and 20% fat and at least 24% protein.
- An antioxidant should be added to reduce oxidation of the fat during storage.
- The fat should be homogenised so that 90% of the fat globules have a diameter of less than 4 microns.
- The milk powder should contain not more than 0.1% crude fibre and the starch content should be stated.
- The proportion of milk protein of the total protein should be stated.
- The milk powder should be supplemented with 6000 IU of vitamin A, 600 IU of vitamin D and 10 mg of vitamin E per kg (IU stands for international units, which are used to measure concentrations of vitamins in feeds).
- The milk powder should contain 100 mg per kg of iron, unless intended for veal production.

More recently, a US organisation called the Bovine Alliance on Management and Nutrition (BAMN) has developed a series of farmer guidelines for calf feeding. Their guidelines on milk replacers (BAMN 1997) uses the following quality parameters.

9.2.1 BAMN guidelines for dry powder

- ***Colour.*** The colour should be cream to light tan, free of lumps and foreign material. If the powder is orange-brown in colour and has a burned or caramelised smell, the product has undergone Maillard browning (non-enzymatic browning) as a result of excessive heat during storage. If the product has 'browned', there will be some loss of nutrient quality and product palatability.
- ***Composition.*** The powder should not contain lumps of CMR or any foreign material.
- ***Odour.*** The powder should have a bland to pleasant odour. A burnt smell indicates heat damage. If it has an odour of paint, grass, clay or petrol, the fat portion of the product may be rancid.

9.2.2 BAMN guidelines for reconstituted liquid

- ***Mixing.*** The product should go into solution easily. Milk replacer should be mixed until all the powder is in solution or suspension without clumps of undissolved powder on the surface of the solution or at the bottom of the bucket. Ingredients that are in suspension, but are not soluble, will settle out of solution (form a sediment) if allowed to stand without agitation. This sediment layer will be more apparent as the fibre content and/or level of added minerals and/or medication increases. In some feeding situations (automatic feeders, nipple bottles, etc.), milk replacers containing significant amounts of insoluble components may not be acceptable. Care should be taken not to over mix. If agitation is continued after the product is in solution, excessive foaming can occur or the fat portion of the product may separate and form a greasy layer on the surface.
- ***Colour.*** The colour should be cream to light tan.
- ***Odour.*** The odour should be pleasant, with no 'off' odours noted.
- ***Flavour.*** The flavour should be milky with no 'off' flavours. Some milk replacers are supplemented with organic acids. These will have a 'tangy' (sweet tart) taste. This should not be confused with a sour taste, which indicates rancid fat.

The best single criterion for evaluating milk replacer is calf performance. If it is poor, more detailed evaluation of management, calf health and milk replacer quality is necessary to determine the reason for the poor performance.

9.3 The nutritive value of milk replacers

The energy content of milk replacers primarily depends on their fat content. The added fat is less digestible than milk fats, so milk replacers generally contain less energy than whole milk supplying the same amount of milk solids.

Formulae are available to calculate the metabolisable energy (ME) contents of milk products and two of these are presented below for the benefit of producers wishing to calculate the energy values and energy costs of the variety of feeds used for rearing calves (Moran 2002).

The ME content of whole milk can be calculated as follows:

$$ME = \frac{[(35.9 \times F) + (19.1 \times P) + 88.8]}{TS}$$

where ME is metabolisable energy in MJ/kg DM of whole milk
F is milk fat (%)
P is milk protein (%)
TS is total milk solids (%).

Table 9.1 lists the ME content of whole milk at various fat, protein and total solid contents. This table presents protein, rather than the solids-not-fat, content because many dairy farmers are now paid on the basis of fat and protein yields. The solids-not-fat content can be converted to protein content by assuming a constant amount of milk lactose and minerals in whole milk, as follows:

$$P = SNF - 5.8$$

where P is milk protein (%)
SNF is solids-not-fat (%)

Table 9.1 shows that the ME content of whole milk can vary from 20 to 26 MJ/kg DM, depending on its composition.

The ME content of milk replacer can be calculated as follows:

$$ME = (0.23 \times F) + (0.06 \times P0 + 14.1$$

where ME is the metabolisable energy in MJ/kg DM
F is the fat percentage in milk replacer DM
P is the protein percentage in milk replacer DM.

Table 9.2 lists the ME content of milk replacer at various fat and protein contents. To allow comparisons with other feeds, these contents are determined on a DM basis whereas the DM content of air dry milk replacer is 96%. This table shows that the ME of commercial milk replacers can vary from 19 to 21 MJ/kg DM, depending on its

Table 9.1. Metabolisable energy content (MJ/kg DM) of whole milk varying in concentrations of fat, protein and total solids (percentage of whole milk)

Total solids	Protein	Fat (%)			
(%)	(%)	3.5	4.0	4.5	5.0
12.5	2.5	21.0	22.5	23.8	25.3
	3.0	21.7	23.2	24.6	26.0
	3.5	22.5	23.9	25.4	26.8
13.0	2.5	20.2	21.6	22.9	24.3
	3.0	20.9	22.3	23.7	25.0
	3.5	21.6	23.0	24.4	25.8

Table 9.2. Metabolisable energy content (MJ/kg DM) of milk replacers varying in concentrations of fat and protein (percentage of DM)

Protein	Fat (%)			
(%)	16	18	20	22
20	19.0	19.4	19.9	20.4
25	19.3	19.7	20.2	20.7
30	19.6	20.0	20.5	21.0

composition. These calculations may underestimate the contribution of lactose to the energy value of milk replacer, particularly in powders with lower than normal fat content.

The nutritive value of milk replacer (of a given composition) relative to that of whole milk (of a given composition) can be calculated by comparing these two tables. Furthermore, these tables can be used to calculate the amount of milk replacer or whole milk required by rapidly growing young calves.

The ME requirements for calves was discussed in Chapter 4 and illustrated in Table 4.1. Milk-based diets are used more efficiently for growth than solid feeds, hence the ME requirements of milk-fed calves are slightly lower than those presented in Table 4.1. For example, 100 kg milk-fed calves growing at 0.5 kg/day each require 21 MJ/day of ME and this is 4 MJ/day of ME less than if they were weaned. For the same growth rate, 50 kg calves each require 15 MJ/day of ME, while 75 kg calves each require 18 MJ/day of ME.

Assuming they are consuming negligible solid food, 50 kg calves growing at 0.5 kg/day while fed milk replacer containing 20% fat and 25% protein (or 20.2 MJ of energy/kg DM), each require 740 g DM/day or 770 g/day of air dry powder. If drinking whole milk containing 4% fat, 3% protein and 13% total solids (or 22.3 MJ of energy/kg DM), each calf requires 670 g milk DM/day or 5.2 L/day. This particular milk replacer then only supplies 91% of the ME for the same amount of DM as this particular whole milk. The daily ME requirements for 75 kg calves growing at 0.5 kg/day would be supplied by 930 g of air dry milk replacer or 6.2 L of whole milk.

9.4 The relative cost of milk replacers

Producers must decide whether to feed CMR or whole milk to their calves. This decision is often based on the relative cost of the two feeds. This can be calculated on the basis of cost for supplying the same total solids (for example, Australian cents per kg DM) or cost for supplying the same feed energy (Australian cents per MJ of ME).

If milk replacer was available for A$65 per 20 kg bag, it would cost 325 c/kg air dry powder or 337 c/kg of powder DM. If it contained 20% fat and 25% protein, it would provide 20.2 MJ of energy per kg DM, and the feed energy supplied would cost 16.7 c/MJ of ME.

Let us assume that whole milk containing 4% fat, 3% protein and 13% total solids, thus providing 22.3 MJ of energy per kg DM, was the alternative liquid feed being considered. Milk replacer would be cheaper only when whole milk cost more than 337 c/kg DM or 43.8 c/L. With milk payments based on milk composition, these calculations become more complex because dairy farmers must consider milk fat, protein and total milk volume. Such calculations are discussed in more detail in Chapter 15.

9.5 Using milk replacers to rear calves

When planning a rearing program based on milk replacer, it is best to order a bulk supply of the replacer because it is often cheaper per bag than smaller lots. Quality control during processing can sometimes be questioned with less well-known brands of replacers, particularly when milk powders become available on the market at extreme discount prices. The generalisation that 'you get what you pay for' holds true for such products. For example, one particular source may be offering cheap CMR because it was subjected to excess heating during processing.

Between 55 and 65% of the total cash costs of replacement heifers is attributable to feed, with much of this occurring post-weaning (Moran 2009). In this context, saving A$5–10/calf on lower cost milk replacers does not seem to be a good economic decision if its poorer quality places the calf at greater risk of nutritional ill health.

It is important that calf rearers understand mixing strengths when preparing milk replacers for feeding. The mixing instructions usually refer to the quantity of powder within a given volume of reconstituted mix, not the amount of water added to the powder. For example, the instructions may be to mix 250 g of powder in warm water and make up to 2 L. If 2 L of warm water were added to the 250 g of powder, the volume of the final mix would be 2.25 L and the calf would have to consume more liquid for the same nutrient intake.

In this first case, making 250 g of powder up to 2 L produces a solution with a strength of 1 in 8 or 12.5%, whereas adding 2 L to 250 g of powder would give a strength of 1 in 9 or 11.1%. The important point is to make sure that the correct amount of milk replacer is measured, or preferably weighed out, for the number of calves being fed. Having a large tank or several very large containers, such as plastic garbage bins, in which to mix the CMR solution makes the job much easier. It is best to have a written recipe for various numbers of calves, so CMR can be mixed for *x* calves simply by adding *y* L of water to *z* kg of CMR powder. This can be made even easier by having standard buckets for measuring the water and converting *y* L of water to *q* buckets of water. If dealing in half or quarter buckets, one of the buckets should be calibrated in half and quarters. A thermometer is also essential to ensure the correct temperature of the final CMR solution. Using hands to estimate temperature of the CMR solution is notoriously inaccurate, particularly on cool mornings.

Once weighed out into a bucket using a spring balance or kitchen scales, the powder should be placed in a calibrated vessel containing some of (but not all) the water, then mixed, either mechanically or by using a hand whisk. The final amount of water is then added to give the correct volume and temperature. It is very important that a consistent feeding temperature be used. For warm solutions, this should be around body temperature: about 36°C, but no more. Some brands of milk replacer can be mixed in cold water and this will be indicated in the instructions written on the bag.

An alternative to weighing is to use a measure, often provided by manufacturers, where one measure of powder is equal to one feed for each calf. In this case, the measure must be regularly checked, because milk replacer powders can vary considerably in bulk density and errors of up to 20% can arise.

Depending on the number of calves being fed, the liquid replacer can be measured out by hand into buckets for individual calves, poured into troughs for communal feeding or into large feeding drums if *ad lib* systems with teat feeding are being used. It can be pumped into individual buckets using a petrol bowser dispenser connected to a large reservoir. With one person feeding, say 50–100 calves, the feeding time will average about half a minute per calf.

The provision of hot water for feeding and washing up afterwards is an important practical consideration. The temperature of cold water can vary from 4°C (in winter) to 15°C (in summer). Heating the water to 70°C and mixing it with tap water, roughly in the ratio of 2:1, produces a final mix of about 40°C. This can be judged by hand, but ideally should be tested each time with a thermometer.

Because milk replacer contains dried milk powders and non-milk products, it behaves differently from fresh whole milk once it enters the abomasum. Curds of whole milk, being more digestible, are broken down more quickly in the abomasum, thus allowing the calf to have more frequent drinks. However, curds of milk replacer must be given more time in the abomasum for their complete digestion.

Milk replacer should be fed less frequently than whole milk. Too frequent feeding of too much milk replacer can lead to abomasal-induced milk bloat. This occurs when the newer clot envelopes the old, partially digested clot of milk replacer, reducing the opportunity for gases to escape and causing distension of the abomasum. It can also lead to overfilling of the abomasum and the spilling over of unclotted replacer into the intestines: a certain cause of calf scours. Twice daily feeding of small quantities of milk replacer can successfully rear calves, but once daily feeding is likely to create fewer problems.

Another role for milk replacers in calf rearing is through boosting the concentration of whole milk. The rationale is that calves can be fed smaller volumes of whole milk, yet consume similar or higher intakes of energy and protein. This would be beneficial to small calves when introduced to once daily feeding. It is essential to provide sufficient drinking water to satisfy the greater thirst of calves when fed whole milk plus milk replacer.

Research with different concentrations of milk replacer have shown that the optimum milk DM concentration for calf growth and feed utilisation is about 15%. Because whole milk contains 12–13% total solids, in theory only 25 or 30 g of powder should be added to each litre of whole milk. However, successful systems have been developed using once daily feeding of 500 g of replacer in 2.5 L of water or of 300 g of replacer in 2 L of colostrum or whole milk.

9.5.1 The final word on milk replacers

- Milk replacers are the substitutes for milk that provide a convenient way to feed pre-ruminant stock.
- They are generally made up of ingredients such as skim milk powder, vegetable or animal fat, buttermilk powder, whey protein, soy lecithin and vitamin-mineral premix.

- A small proportion of other ingredients like glucose, non-milk protein and cereal flour can also be used.
- Pre-ruminant stock of less than 3 weeks of age should preferably be on milk replacer made of all milk protein.
- Milk replacers can be stored long term as powder and reconstituted by mixing with water as recommended.
- Young stock can grow equally well when reared on milk replacer and their rumens can develop just as well as they would on a diet of whole milk.
- The cost of milk replacer must be competitive with whole milk for livestock farmers to consider using it when they may have copious supplies of fresh milk from their own farm.
- In a mixed farming system, surplus fresh milk from one ruminant species can be reconstituted to formulate milk replacer for another ruminant species.
- If by-products of whole milk, such as skim milk or whey, are produced on farm they can also be used as a partial replacer for whole milk or can be reconstituted as a milk replacer with the addition of required ingredients.
- Despite these options, most farmers in developing countries only use whole milk in their pre-ruminant rearing systems.
- By increasing awareness about the potential of preparing milk replacers and feeding to pre-ruminant young stock, this will lead to improved survivability and growth of pre-ruminant stock.

10

Solid feeds for milk-fed calves

This chapter discusses the ingredients and the formulation of solid feeds for milk-fed calves.

The main points in this chapter

- Calves have high requirements for energy and protein. When purchasing calf-rearing concentrates, they must be formulated for the calves' requirements and not for those of milking cows.
- When formulating such a ration, it is important to obtain accurate analyses of energy and protein contents of the ingredients.
- Concentrates and drinking water should be made available from the first week of milk rearing. Milk-fed calves should be offered fresh concentrates every day.
- Ideally, limited roughages should be offered during the milk-rearing phase to stimulate rumen development.
- Calves can be successfully weaned off milk when consuming 0.75–1.0 kg/calf/day of concentrate.
- Weaned calves should weigh at least 70 kg and be seen to be ruminating.
- The 12-week live weight for Friesians can vary from 85 to 125 kg, depending on milk intake and the success of the transition phase from milk to solid feeds.
- A realistic target for Friesian heifers is 95–105 kg at 12 weeks of age.
- Specially formulated calf concentrates are not readily available in many Asian countries, so farmers should enrich existing cow milking concentrates with additional protein.

This chapter deals with the nutritive value of solid feeds and their formulation into diets for milk-fed calves. Calves require feeds rich in energy to promote high feed intakes and good animal performance. The energy value of feeds is measured in terms of their metabolisable energy or ME concentration (as discussed in Chapter 4). Feed energy is usually provided by carbohydrates, which come essentially in two forms: material inside plant cells such as starch (in cereal grains) and sugars (in high-quality pastures), and digestible material in the cell walls such as cellulose. As their rumens develop, calves can

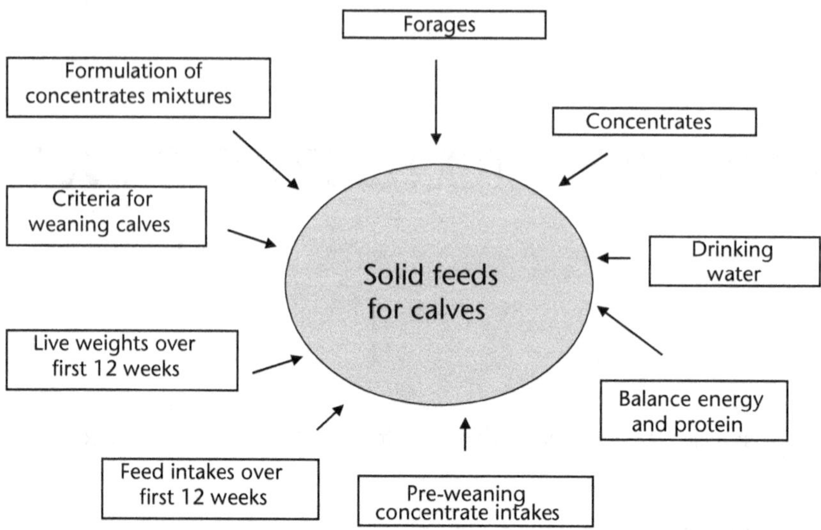

Figure 10.1. Decisions to make when considering solid feeds for calves

begin to digest the energy themselves, but require the rumen microbes to digest and use the plant cell walls.

Calves and heifers also require feeds that are high in crude protein (CP) and supply undegradable dietary protein (UDP) as well as rumen degradable protein (RDP). RDP originates from the feed nitrogen – both true protein and non-protein nitrogen – which is broken down in the rumen into ammonia to provide one of the basic nutrients for rumen microbes to grow and produce microbial protein. This microbial protein together with any feed protein escaping rumen digestion (the UDP) then passes into the abomasum for digestion by the calf's own gut. These processes have been described in more detail in Chapter 4. The key decisions to make when considering solids feeds are summarised in Figure 10.1.

10.1 The nutritive value of solid feeds

Most dairy cattle nutritional reference books contain tables of nutritive value of feeds and supplements. The following two tables provide a summary of some of the common forages (Table 10.1) and supplements (Table 10.2) fed to dairy stock in Asia (Moran 2005). These tables describe feeds on the basis of their ME, CP and neutral detergent fibre (NDF) contents. The values are in ranges to take into account differences in the agronomic growing conditions and stage of maturity in forages and the degree of processing of by-products.

Table 10.3 classifies feeds on the basis of their average CP and ME contents thus provides a ready reckoner to consider when seeking feeds to balance heifer diets that may be low in dietary energy or protein.

The rumen capacity of weaned heifers does not reach mature proportions until 5–6 months of age and, unless pasture quality is high, at least 10 MJ/kg DM, feed intakes and growth rates are restricted by limited rumen volume. With heifers weighing less

Table 10.1. Nutritive values of some common Asian forages

Forage	DM (%)	ME (MJ/kg DM)	CP (%)	NDF (%)
Grasses or legumes				
Napier grass (immature)	13–15	8–9	11–15	60–70
Napier grass	18–25	7–9	10–12	65–75
Rhodes grass	20–25	7–9	8–10	65–75
Guinea grass	20–25	7–9	10–12	65–75
Para grass	25–30	7–9	10–12	65–75
Forage sorghum	13–15	7–9	11–15	65–75
Native pasture	25–35	7–9	8–10	60–70
Leucaena leaf	28–32	7–9	18–26	40–50
By-products				
Maize stover	18–25	7–8	6–8	60–70
Cassava hay	85–90	8–9	22–25	50–60
Peanut leaf and stem	26–30	9–10	18–20	50–60
Soybean leaf	26–30	10–11	20–24	50–60

Table 10.2. Nutritive value of some common Asian supplements

Feed	DM (%)	ME (MJ/kg DM)	CP (%)	NDF (%)
Energy				
Cassava chips	85–90	12–13	2–3	5–10
Maize grain	85–90	13–14	10–12	8–10
Rice bran (good quality)	85–90	11–12	13–14	25–35
Rice bran (poor quality)	85–90	8–10	8–10	30–50
Sweet potatoes	30–35	12–13	5–6	20–30
Cassava waste (fresh)	20–24	7–8	2–3	20–25
Wheat pollard	85–90	11–12	14–16	30–40
Brewers grain	25–30	10–11	25–30	50–60
Commercial concentrate	85–90	11–12	15–18	12–25
Protein				
Soybean meal	85–90	13–14	45–50	25–35
Coconut meal	85–90	12–13	15–25	40–50
Cottonseed meal	85–90	12–13	40–45	30–40
Palm kernel cake	85–90	11–12	14–16	60–70
Soybean curd (fresh)	13–15	13–14	20–25	25–35
Urea	100	–	280	–
Forage				
Rice straw	85–90	6–7	4–6	65–75
Urea-treated rice straw	85–90	8–9	7–8	65–75
Banana stem	5–8	9–10	3–4	60–70
Sugar cane tops	25–30	7–9	5–6	65–75

Table 10.3. Classification of Asian supplements and basal forages according to their energy and protein contents

Energy/protein classification	Poor energy (<8 MJ/kg DM of ME)	Moderate energy (8–10 MJ/kg DM of ME)	Good energy (>10 MJ/kg DM of ME)
Poor protein (<10% CP)	Rice straw Maize stover Sugar cane tops Cassava waste	Rice bran (poor) Most grasses Sweet corn cobs Banana stem Urea-treated rice straw	Cassava chips Paddy rice Molasses Sweet potatoes Pineapple waste Maize silage
Moderate protein (10–16% CP)	–	Brown rice Well-managed grasses Soybean Immature grasses	Maize grain Sorghum grain Rice bran (good) Wheat pollard Palm kernel cake
Good protein (<16% CP)	Urea	Whole cottonseed Shrimp waste Cassava hay Most legumes Legume hays	Brewers grain Coconut meal Soybean curd Commercial concentrate Protein meals Legume leaves

than 200 kg, dietary fibre content has a greater influence on intake and growth than energy content, whereas above 200 kg, heifer performance is less affected by ration fill characteristics. Consequently, high energy supplements are usually required to maintain good growth rates in young weaned heifers.

Even when at their best, tropical pastures fall short of being the complete feed for young stock under 6 months of age. Multiple suckled calves can balance the deficiencies in grass by frequent drinks of milk and thus sustain high growth rates. However, even with limited access to milk, growth rates of young calves fed pasture alone suffer from restricted nutrient intakes, particularly dietary energy. This will be discussed later in this chapter.

The three tables of feed quality highlight the wide variation that can occur in energy and protein levels within any particular type of feed. When formulating rations for young stock, or for any livestock, estimates of the quality of the diet are only as good as the information available on the ingredients. It is important that accurate values of energy and protein contents of the feeds actually being fed are used in ration formulation.

10.2 Feed intake and calf performance pre-weaning

The contribution of solid feeds to the performance of young calves fed limited milk or milk replacer can be quite large, particularly when they are weaned at very early ages. For example, in a survey of 30 calf feeding trials in the US with weaning ages averaging 32 days (ranging from 19 to 45 days), calves averaged 0.3 kg/day of concentrates (ranging from 0.1 to 0.5 kg/day) and grew at 0.3 kg/day prior to weaning (ranging from 0.1 to 0.5 kg/day).

Table 10.4 summarises data on concentrate intakes and growth in calves fed 450 g of milk replacer in 3.8 L of water each day for 21 days and then half this amount until

Calves can be fed limited amounts of clean and palatable hay prior to weaning.

weaning on day 28. Roughages were not fed during this period. The table presents average values each week together with the range for poor to good calves.

10.3 Feed intake and calf performance throughout the rearing period

Once calves are successfully weaned, concentrate intake rapidly increases. Figures 10.2, 10.3 and 10.4 show concentrate intakes and growth rates in Friesian bull calves reared indoors on either *ad lib* or limited milk replacer from week 1, when they are bought at

Table 10.4. Concentrate intakes and growth rates in pre-weaned calves

Age (weeks)	Concentrate intake (kg DM/day)		Growth rate (kg/day)	
	Mean	Range	Mean	Range
1	0.1	0–0.1	0.1	0.1–0.2
2	0.2	0.2–0.3	0.1	0.1–0.2
3	0.5	0.4–0.7	0.5	0.4–0.7
4	1.0	0.9–1.2	0.6	0.5–0.7
Average for 4 weeks	0.5	0.4–0.6	0.3	0.3–0.4

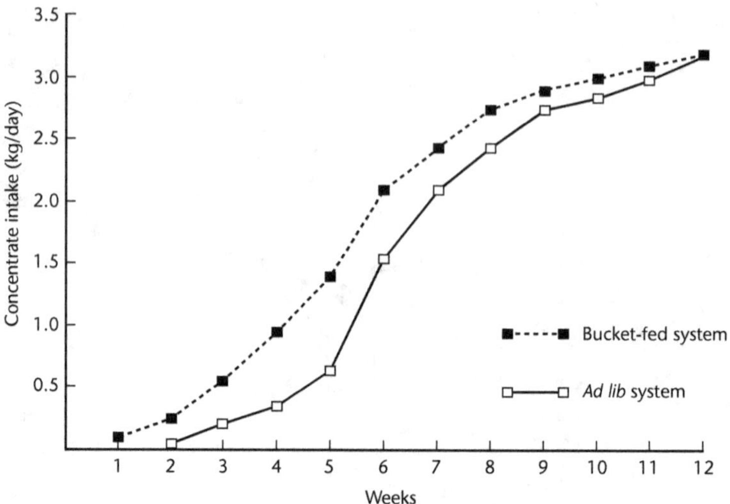

Figure 10.2. Daily intakes of concentrates when fed *ad lib* from week 1 to week 12 in Friesian bull calves fed either *ad lib* or limited milk replacer until week 5

about 10 days of age. The calves fed *ad lib* each consumed 30 kg milk replacer; those on limited once-daily feeding only consumed 12 kg.

Concentrate intakes increased rapidly following weaning at 5 weeks, particularly in calves previously fed *ad lib* milk replacer. By 12 weeks, both groups of calves were eating similar amounts of concentrates. Daily intakes are shown in Figure 10.2 and cumulative intakes in Figure 10.3. Each animal required 130–150 kg of concentrates over the full 12-week period but this would be reduced to 100 kg or less if they were grazed by say, 8–10 weeks of age.

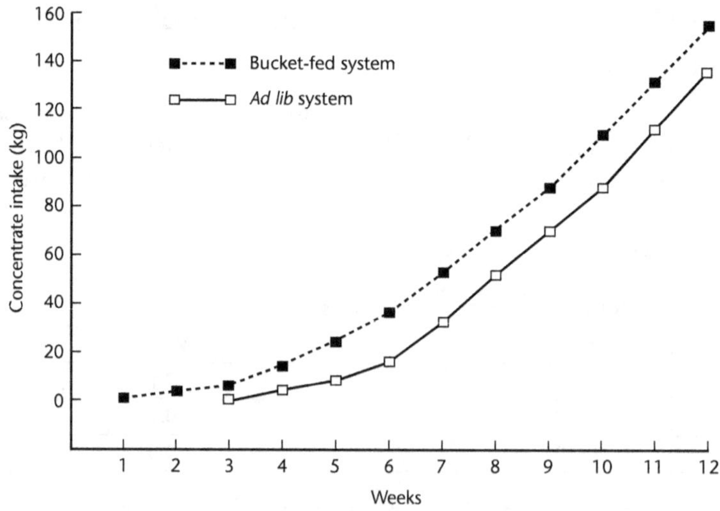

Figure 10.3. Cumulative intakes of concentrates when fed *ad lib* from week 1 to week 12 in Friesian bull calves fed either *ad lib* or limited milk replacer until week 5

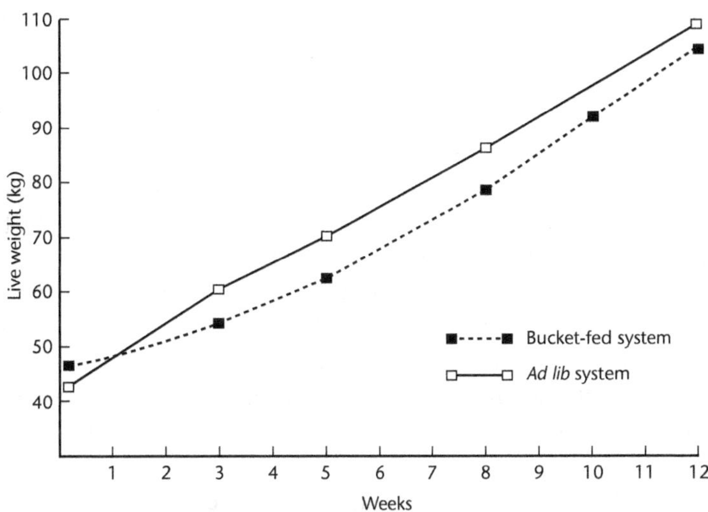

Figure 10.4. Live weights in Friesian bull calves fed *ad lib* or limited milk replacer until week 5 together with *ad lib* concentrates from weeks 1 to 12

Figure 10.4 shows live weights of calves in the two rearing systems. It is of interest that the extra gain made by the *ad lib* fed calves was all achieved in the first 3 weeks of rearing. One useful measure of quality of management of calf-rearing systems is calf live weight at 12 weeks of age, which would normally mean after 11 weeks of rearing. This single measure takes into account the feeding and management during milk rearing, weaning and early post-weaning growth.

The 12-week weight in calves can vary from 85 kg or less to more than 125 kg, depending mainly on milk intake and the success of the transition phase from milk to solid feeds. Live weights of 95–105 kg at 12 weeks of age would indicate a well-managed calf-rearing system.

10.4 Criteria for weaning calves

Different fashions in calf husbandry in different areas of the world have tended to favour different ages as optimal for weaning dairy calves, from as early as 4 to as late as 12 weeks after birth. However, the trend over the last 20 yr has been towards weaning as early as possible. There are several good reasons for this:

- The feed energy in whole milk or CMR costs up to four times more than the feed energy in concentrates and up to 20 times more than the energy in grazed pasture or hand-harvested forages.
- Liquid feeding is very labour intensive and time consuming.
- Facilities for rearing calves from birth to weaning, such as pens, are more costly than those required after weaning and the shorter the period, the fewer the pens required.
- Disease control, particularly scours, is easier to manage in weaned calves.

One of the most important practical considerations in the pre-weaning period is to ensure that the calf stays alive and that it does not succumb to disease severe enough to set back its growth. Therefore the most economical feeding system prior to weaning may not necessarily be the one that costs the least in food and labour.

Although the rumen of a 3–4-week-old calf may be as effective as that of an adult animal, the rumen capacity should be the major determinant for weaning. This depends on dry feed, hence (indirectly) on milk intake. For example, an aggressive calf that drinks more milk than its pen-mates will probably eat less concentrates and roughage. At the same age, it may be heavier than its pen-mates, but its rumen will be less developed. Therefore, this particular animal should be weaned at an older age.

Calves can be successfully weaned onto dry feed when eating 0.5 kg/day of concentrates. This limit should be increased to 0.75–1.0 kg/day, because farmers do not want their calves to have any post-weaning setbacks. Individual concentrate intakes are difficult to estimate in group-housed animals. However, this level of intake normally occurs around 6 weeks of age. Weaned calves should weigh at least 70 kg and be seen ruminating.

The bigger the calf when entering the rearing unit, the quicker it can be weaned. For every 10 kg increase in initial live weight, it should take 7 days less to reach the same intake of concentrate.

10.5 Tips to stimulate concentrate intakes

There are several management strategies available to stimulate calf starter intakes in milk-fed calves. These include:

- Ensure that calves have access to high-quality starter.
- Provide adequate clean water.
- Increase the amount of feed offered to ensure it is continually available.
- Reduce the amount of fine particles that separate out from pelleted calf starter.
- Calves reared in groups learn to eat calf starter earlier than individually reared calves.
- Calves with good immunity levels, arising from good colostrum feeding management, eat calf starter earlier than those with failure of passive transfer.
- 'Step-down' milk feeding or reducing the quantity of milk or CMR offered at say 3–4 weeks of age, can stimulate starter intake.
- Feeding long hay can stimulate starter intake, but not always.
- Minimising heat stress, if it is a problem, will increase starter intake.
- Minimising any negative effects of birds, flies and other pests on starter quality, pen hygiene and calf health, will increase starter intake.

10.5.1 Palatability

One important factor with calf concentrates is palatability: calves must like to eat it. Palatability is influenced by several factors such as flavour, texture and ingredients. Flavour is most often added by including molasses, which can be included in pellets or spraying on a textured feed. The amount of molasses influences handling, and adding too much can cause the feed to stick together in hard clumps, especially in cold weather. Mould and staleness contribute to unacceptable flavours. Certain ingredients can adversely influence flavour such as high levels of additives like urea, sodium bicarbonate or ionophores (compounds used to improve the efficiency of rumen digestion).

Texture is important because calves prefer textured feed to ground meals. Powdery or dusty feeds are less palatable and these can be found in feeds with poorly formed pellets which contain a lot of 'fines' or small particles that separate out of the pellets.

10.6 Formulating concentrates for weaned calves

The first concentrate mixes offered to milk-fed calves are often called starter rations. It is becoming more common for farmers to use commercially produced pellets rather than mix them from raw ingredients on farm. This is because they must be highly palatable, fresh when fed and specifically formulated to provide the correct balance of nutrients for the transition period from milk to solid feeds.

These rations should contain at least 18% crude protein and 12 MJ/kg of ME. The inclusion of rumen buffers, such as sodium bicarbonate, in calf starters has also been shown to improve intakes and growth rates. They are frequently packaged in small-sized (3–5 mm) pellets. This reduces dust and ensures that calves cannot select out any ingredients they like and reject others.

Several practises are used to encourage very young calves to nibble starter rations at an early age. These include placing a small amount of pellets in each milk bucket, assuming this is the milk feeding system used, as they finish their daily allocation of milk. Calves may not initially eat more pellets but intakes can increase in later weeks by over 25% and this has been shown to improve growth rates to 4 weeks by 40%.

Young calves should only have fresh pellets available. They should be offered a new batch of pellets each day, while those left from the previous day can be fed to older weaned calves. Twice daily feeding of concentrates is often recommended to give the slower eating calves a better chance of having their full daily ration. Fresh water must be available from the start.

The use of ground forage or coarse grain materials (such as husks) in calf diets has had limited research attention. Solid feed intake and live weight gains have been found to be greater in calves offered textured feed, intermediate for ground feed and lowest for pelleted feeds (Margerison and Downey 2005). Calves have also benefited from including 10–15% of their solid feeds as chopped hay, but this would pose problems for manufacturing and packaging of calf concentrate formulations.

One problem with milk rearing calves at pasture is encouraging them to eat starter pellets at an early age. Calves seem to prefer fresh grass (particularly in high-quality spring pasture) to pellets. Feeding the starter rations in textured form, by including molasses, flaked as well as rolled cereal grains, together with some coarsely chopped, highly palatable hay, may be one way of overcoming low concentrate intakes in very young calves reared at pasture.

When starter rations are combined with straw, penned calves eat more pellets because of a more stable level of acidity in the rumen. Research in Australia has shown that giving young calves access to straw increased their pellet intakes by 15% and their growth rates by 25%. Providing calves with better quality hay will increase roughage intake at the expense of the pellets and so reduce growth rates. The better the quality of roughage, the less pellets eaten. Other forms of roughage, such as cottonseed or oat hulls, could be used.

It must be emphasised that to be effective, the roughage must be coarsely chopped and not finely ground or milled before inclusion in the pellets. Excellent calf performance has been achieved by grinding roughage through a 22 mm screen (10 mesh) and incorporating the complete diet in 5 mm diameter pellets. The handling of chopped straw or other fibrous by-products by feed companies would require additional processing and equipment.

Growing heifers need to be supplemented with the correct mineral and vitamin package. Calcium and phosphorus are vital for the development of large-framed animals with strong healthy bones. Phosphorus also promotes feed intake. Minerals are important nutrients that can affect fertility and hence the ability of heifers to calve down within 24–30 months of age. The minerals most associated with infertility include copper, zinc, selenium, iodine and phosphorus.

10.6.1 Sourcing calf concentrates in Asia

Throughout Asia, the majority of formulated supplements for milking cows are formulated to 16% CP, even though on closer investigation (Moran 2005), they are frequently below this content. For convenience, many SHD farmers also feed these concentrates to their young stock. Such formulations are far from ideal because, for optimal growth and health, milk-fed calves and weaned heifers require 18% CP in their total diet (basal forage plus supplements). Depending on the quality of the basal roughage fed post-weaning, 18% may be insufficient for the concentrates. The only alternative is then to enrich the available cow milking concentrates with additional protein to increase them to at least 18–20%. These concentrates should not include urea because milk-fed calves cannot use non-protein nitrogen.

Very rarely can small holder farmers purchase higher protein formulated concentrates and in many cases they are not even aware of the benefits for their young stock in supplementing available milking concentrates with additional protein supplements.

High protein concentrates may be available, but at great expense, because they have been formulated for pig and poultry, incorporating high-quality protein ingredients. It would be ideal if a few large-scale feed mills, either owned by dairy cooperatives or agribusiness, could formulate calf and heifer mixes with higher protein contents, using better quality energy sources and additional minerals and vitamins for optimal growth of young stock. Compared with the higher demand of concentrates specially formulated for milking cows, the formulation of smaller batches of calf/heifer mixes would not be cost effective for small dairy cooperatives. Specially formulated calf concentrates are becoming available in some SE Asian countries, such as Malaysia. The economic benefits of such feeds will be discussed in Chapter 15.

11

Disease prevention in calves

This chapter describes the clinical symptoms of the major calf diseases and the first aid and nursing during the calf's sickness and convalescence.

The main points in this chapter

- There will invariably be some calves born dead or will die pre-weaning. In developed temperate dairy systems, excluding abortions, 7–9% of calves are expected to die between birth and 3 months of age, although when well managed, the norm is 2–4%. However, a range of 15–25% pre-weaning mortality would be typical on many tropical dairy farms.
- The best way to maintain calf health is to ensure an adequate intake of colostrum immunoglobulins within the first few hours of life. Prevention through adequate colostrum intake is far more effective than cure by drugs.
- The two major diseases of calves are scours and pneumonia, which account for 80% of all calf deaths. Bloat, navel-ill, accidents and poisoning make up the bulk of remaining mortalities.
- The cause of scours in calves under 21 days of age is difficult to determine. There is usually not one single cause, but an interaction between calf management, diet, the environment, poor immunity, and the presence of pathogenic viruses and bacteria.
- Most of the scours can be controlled through good management and appropriate preventative measures.
- Pneumonia is a problem with housed calves, particularly when stocking density is high and ventilation is poor. Control is mainly through improved housing.
- Clostridial diseases, such as pulpy kidney, can be easily prevented through a routine vaccination program.
- Internal parasites are less of a problem with housed than grazed calves.
- Sick calves can be most easily recognised through changes in behaviour. They are more likely to be culled for poor performance later in life. Keeping records will assist with decision making on their future.
- In the event of a veterinary visit, there is much the dairy farmer can do to prepare for it and to care for the convalescing calf.
- Because calves are the class of dairy stock most susceptible to diseases, every effort should be made to maintain a healthy shed environment. Developing an effective biosecurity program, which restricts high-risk visitors, is an integral part of good calf management.

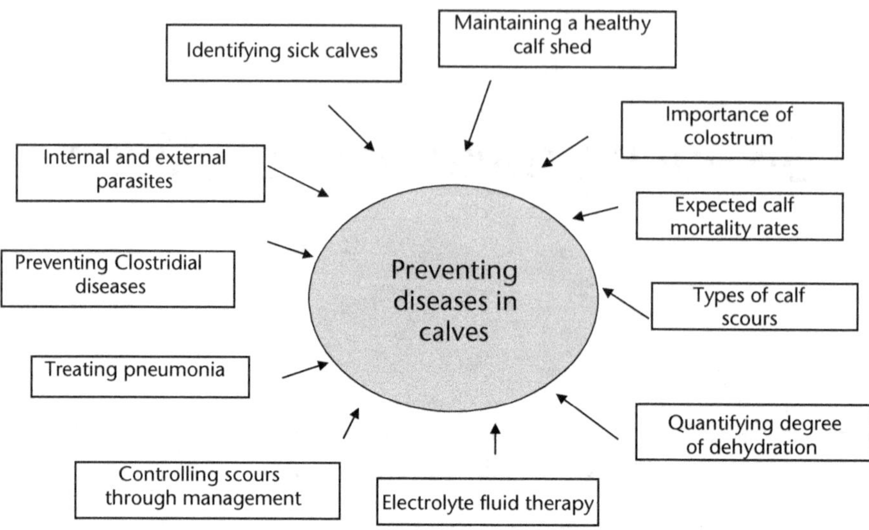

Figure 11.1 Some of the key factors to consider to control and prevent calf diseases

This chapter concentrates on the clinical signs of the major calf diseases and on first aid and nursing during sickness and convalescence; many of these are summarised in Figure 11.1. This chapter does not present a comprehensive catalogue of calf diseases, nor does it follow the pursuit of a diagnosis through post-mortem examination, microbiology and clinical pathology.

There will invariably be calves that are either born dead or die pre-weaning. What constitutes an acceptable death rate? Roy (1990) considers that under good management in temperate, developed dairy industries, expected mortality rates are:

- abortions (stillbirths <270 day gestation): 2–2.5%
- peri-natal (stillbirths >270 day and during first 24 hr of life): 3.5–5%
- neonatal (between 24 hr and 28 days of life): 3%
- older (29–84 days) 1% or (84–182 days): 1%.

So excluding abortions, 7–9% of calves often die between birth and 3 months of age. This seems to be the reported rate in the US, whereas in Australia it is generally lower, say at 2–4%. As discussed in Chapter 6, a range of 15–25% pre-weaning mortality would be typical on many tropical dairy farms.

The basic principles for good calf health are:

- Minimise exposure to disease pathogens.
- Delay any exposure until calves can develop their own immunity.
- Maximise acquired immunity through colostrum and vaccinations.
- Keep calves sheltered, dry and free from stress.

The best way to maintain calf health is to ensure an adequate intake of colostral antibodies within the first few hours of life. Good farm management should ensure this occurs (see Chapter 5). This is obviously difficult if relying on calf purchases to supply animals for rearing. If calves have to be bought, it is preferable to buy them directly from the property of origin because this reduces their likelihood of picking up diseases in transit and the duration of stress and starvation. Prevention by adequate colostrum intake is far more effective than cure by drugs.

Rearing calves inside sheds at high stocking densities can provide an ideal environment for calf diseases to proliferate, although many still occur in calves reared at pasture. Prevention of future outbreaks through cleaning and disinfection is also more difficult when calves are reared in permanent fixtures. However, a warm, dry and well-managed calf shed usually offers better protection against diseases in young calves, particularly during cool weather, than any cold, windy and muddy calf paddock. Well-ventilated sheds with auxiliary cooling (using fans and sprinklers) provide better climate control in hot weather than exposed paddocks.

There are two major disease problems in calves in Asia: namely scours and pneumonia. These two would account for more than 80% of all calf deaths, with scouring being the most common. Bloat, navel-ill, accidents and poisoning would make up the bulk of the remaining mortalities.

11.1 Calf scours or neonatal diarrhoea

Normal faeces has one colour and consistency when the calf defecates. In milk-fed calves, it usually is dark yellow in colour, but a lot of variation from this is normal. When exposed to air, the faeces becomes darker and more solid. In calves eating some concentrates, normal faeces will be browner and slightly more solid. Changes in colour and/or consistency can indicate scours. However, the colour and odour of the faeces cannot positively identify the invading organisms causing scours.

The causes of scours in calves under 21 days of age are difficult to determine. There is usually not one single cause but an interaction between calf management, diet, the environment, poor immunity, and pathogenic viruses and bacteria.

11.1.1 Types of scours

Dietary scours

This mainly results from overfeeding (especially with cold milk) or incorrect milk replacer concentrations. Sudden changes in feed type – particularly changing from whole milk to milk replacer, or use of poor-quality milk replacers – can also lead to dietary scours. Affected calves get severe diarrhoea but otherwise appear normal. However, they can more easily develop infectious scours. The best control measure for dietary scours is changing from milk to electrolytes for at least 24 hr. Some farmers and experts recommend taking calves off milk only as a last resort and then only after they are certain that an infective agent is the major cause of scours.

White scours

This generally occurs in the first few days and is usually caused by pathogenic strains of bacteria known as *Escherichia coli* (or *E. coli*), which invade the gut wall. Foul-smelling, grey to creamy-white, severe diarrhoea is seen. Calves quickly become dehydrated and lethargic, will not eat, are 'tucked up' in the abdomen and may die suddenly. In chronic cases that linger on, infection of the lungs (pneumonia) or joints (arthritis) can occur. On post mortem, a calf that died from *E. coli* scours will often show no visible signs of having an infection. Stress factors, such as cold or partial starvation (due to irregular feeding intervals as occurs when calves are purchased through saleyards), can increase the occurrence and severity of white scours.

Viral and protozoal scours
These are generally caused by rotavirus or coronavirus (viral) or *Cryptosporidia* (protozoal) and constitute most of the scours in calves less than 3 weeks old. Antibiotics do not kill viruses or protozoa and so are not effective in treating these scours. Furthermore, their overuse in treating scours will increase the risk of antibacterial residues in slaughtered bobby calves.

***Salmonella* scours**
This occurs more commonly in older calves causing bloody, putrid diarrhoea containing mucus. They develop fever, are weak and rapidly become dehydrated and emaciated. They have a high death rate. Less severely affected calves can have rough coats, pot bellies and become stunted; they can also become carriers of *Salmonella* and continually infect other animals. Extra personal hygiene is needed when treating *Salmonella* because the bacteria can be passed onto humans.

Worm scours
These are caused by internal parasites eaten by grazing calves. These would not occur in housed systems unless purchased calves are older and have previously run at pasture. See Section 11.4 on internal parasite control.

Coccidiosis or blood scours
This is caused by protozoa infecting the calf from 3 weeks of age and onwards and can easily be confused with white scours. Affected calves show blood-stained scouring with a lot of mucus and may eventually develop anaemia. Coccidiosis is a stress-related disease and usually affects calves that are reared in wet, crowded and unhygienic conditions.

Scours accounts for 75% of all deaths under three weeks of age. The most important pathogens associated with infectious scours at different ages are:

- *E. coli*: 3–5 days
- rotavirus: 7–10 days
- coronavirus: 7–15 days
- *Cryptosporidia*: 15–35 days
- *Salmonella*: several weeks
- *Coccidia*: older than 3 weeks.

11.1.2 Treating scours

Scouring calves can lose up to 20 times more fluid than healthy animals and they will become dehydrated because they are losing considerably more liquid than they can

Table 11.1. Measures of dehydration in scouring calves

Percentage dehydration	Sunken eyes*	Skin fold test (seconds)	Clinical symptoms
4–6	–	1–2	Mild depression, decreased urine output
6–8	+	2–4	Dry mouth and nose, tight skin, still standing
8–10	++	6–10	Cold ears, unable to stand
10–12	+++	20–45	Near death

* The more +s, the more sunken the eyes.

drink. This lost fluid also contains mineral salts and other nutrients. The degree of dehydration can be assessed using the skin fold (pinch) test. Pinch the skin and note how long it takes to return to normal: in healthy calves this is less than half a second. Another indicator is the degree of sunkenness of the eyes. Table 11.1 provides some visual indicators of the degree of dehydration.

Very dehydrated calves (10–15%) will require intravenous therapy. Calves with less than 8% dehydration and still drinking can be rehydrated orally by electrolyte solutions. Oral fluid therapy is the term used for treating scours with soluble sources of energy and electrolytes by mouth. These supply an energy supplement and replace lost vital minerals and fluids in scouring calves.

The amount of fluid required for daily maintenance requirements and to replace lost fluids can be calculated, based on live weight and the degree of dehydration. For a 40 kg calf with 6% dehydration:

- replacement: 40 kg × 6% or 2.4 L of fluid
- maintenance: 100 mL/kg/day, or 40 kg × 100 mL or 4.0 L of fluid
- total: 2.4 + 4.0 or 6.4 L of fluid. Feed this quantity in three feeds per day, and check the degree of rehydration using the skin fold test.

Up to 70% of calves will recover with adequate fluid therapy. Electrolyte treatments do not provide sufficient energy to maintain the animal. After 24 hr, reintroduce milk (if it has been withdrawn), but continue electrolytes for a further 48 hr. Separate milk feeding from electrolyte feeding by 6 hr. Rennet (junket) tablets added to the first two milk feeds will help clot the milk hence assist with its digestion.

The electrolyte solution should be offered to calves in the same manner as their milk (bucket or teat), but, if they do not drink it this way, it can be administered using a drench gun or a stomach tube. It is preferable to ask the veterinarian to give intravenous fluids to very sick and dehydrated calves because force feeding often results in pneumonia because such weak calves cannot swallow properly.

Prolonged use of antibiotics can lead to additional scouring because normal bacteria have been killed. This is known as medicine disease. Dosing the calf with plain non-pasteurised yoghurt helps re-establish the abomasal bacteria (*Lactobacillus*) used in milk digestion.

Veterinary advice should be sought to obtain an accurate diagnosis and the most appropriate treatment. Sick calves should be isolated from healthy calves and tended to after feeding other calves to minimise the spread of infection. Drinking water should be freely available.

Diarrhoea powders containing kaolin, pectin or chalk or other methods of slowing down feed passage through the gut (such as charcoal tablets, cornflour or even sawdust) can reduce the severity of the scouring. Antibiotics may be required, especially if the calf remains dull after rehydration, and if blood appears in the faeces. Antibiotics must be used under veterinary supervision. Antibacterial compounds and antibiotics (for example calf scour tablets, drenches or injections) should be used judiciously and restricted to cases where *Salmonella* or other bacteria are suspected.

When prescribed, antibiotics are usually given orally for about 3 days. If the scouring is too advanced and the gut wall is badly damaged or the calf is running a temperature, a course of antibiotic injections may be required.

The infective organisms may be resistant to many disinfectants and survive in the environment for long periods. Formalin and hypochlorite are probably the most effective disinfectants, but only on well-cleaned floors and surfaces. Paddocks and yards are impossible to disinfect and require prolonged spelling. If possible, change the calf-rearing area regularly because the risk of the disease is related to the build up of organisms. This is obviously easier if calves are reared outside in paddocks.

11.1.3 Controlling scours through management

Dietary scours (also known as nutritional scours) is caused by stresses reducing the production of digestive acids in the abomasum. Pathogens consumed by the calf are normally killed by the very low pH from these digestive acids. If the acid production is reduced then the abomasum does not protect the calf from these pathogens and they pass through into the intestines. The low acid production also reduces the effectiveness of the rennet in clotting the milk into a curd and so undigested milk then escapes into the intestines where it cannot be properly digested in the alkaline environment.

Because the bacteria that normally reside in the intestines now have a new supply of nutrients, they multiply and irritate the gut wall. This causes the body to secrete fluids into the intestines, thus leading to a loss of valuable minerals. Hence, a scouring calf becomes rapidly dehydrated and deficient in minerals. If dietary scours is not corrected promptly, the pathogenic bacteria that were not killed off by the stomach acids will also multiply in the undigested milk and the calf will develop infectious scours. By removing the initial stress, sufficient abomasal acids are produced, and normal milk digestion will eventually resume. Sudden changes in milk feeding routines are a common cause of scours. For example, calf rearers routinely report scours in calves about a week after changing from whole milk to milk replacer.

Environmental stress is another cause, such as sudden changes in weather or cold, damp, draughty or humid conditions inside calf sheds. Overcrowding is another cause, so sheds should never house more calves than they were designed to. Even changes in staff can lead to scours through different handling of calves, lack of tender loving care (TLC) or changes in standards of hygiene. If reared outdoors, calves should always be offered protection against the extremes of sun, wind and rain. Despite this precaution, a sudden cold and wet spell can introduce sufficient stress to increase the incidence of scours in well-managed calves.

The duration of scours is largely under the control of the calf rearer. During their second week of life, calves are particularly susceptible. By careful observation, experienced rearers can anticipate the onset of scours the day before it happens, after which milk feeding can be reduced, with the calf recovering quickly.

The following signs of impending scours should be looked for:

- dry muzzle
- thick mucus appearing from the nostrils
- very firm faeces
- refusal to drink milk
- a tendency to lie down
- a high body temperature (over 39.2°C).

Scours can occur under the best management but some precautions always help. If using a calving pad for calving down cows, calves should be quickly removed from any area used for holding these cows prior to calving to reduce the chances of manure contamination of newborn calves. This is also important in the prevention of navel infections and Johne's disease (see Section 11.5).

A feeding routine should be quickly established, with set feeding times, constant amounts of milk offered (and drunk) per calf and a consistent milk temperature. Any changes in feeding routine should not be too sudden. It is best to pen newly purchased calves individually for the first 2 weeks, particularly if they are obtained from various sources, to quarantine them against spread of disease to other calves. If buying calves from selected farmers, ensure that these farms give their newborn calves adequate colostrum, have a low level of calf scours and good management immediately following birth. Milk feeding equipment should be thoroughly washed and sanitised between feeds.

It must be kept in mind that in many (if not most) cases, the causes of scours are many. This means that it is due to more than just one type of pathogen, with environmental factors, nutrition and management all possibly contributing to the infection. However, early identification and treatment of sick calves is the key to their rapid return to health. In fact, many scouring calves that are quickly identified, diagnosed and treated can return to a more normal health state after just several days on fluid replacer treatment when they can be gradually reintroduced to their normal milk feeding routine. Delaying treatment can complicate the disease condition because the weakened, hence more susceptible calves, can quickly succumb to other contributory factors to their infectious scours.

Some calf rearers routinely include small amounts of disinfectant or antibiotics in all milk fed to young calves. This is not good rearing management because it may lead to low levels of infection in all animals, with the infection only apparent when these calves develop more advanced symptoms of the disease. This practice can also increase the growth of antibiotic-resistant organisms, so making it even harder to treat sick calves.

Furthermore, many calf rearers routinely used antibiotics to control any potential pathogens, as well as to increase feed intake and utilisation. This is not necessary with ideal management and facilities, such as where colostrum intake is adequate, the rearing unit is clean and well ventilated and not densely stocked and the operator is experienced. Because this ideal scenario is not common, antibiotics have been used as insurance against disease, particularly when rearing calves bought from often unknown sources. This could mask any disease outbreak for several days and also give a false sense of security, which often leads to an even poorer job in calf raising. Concern about the development of antibiotic-resistant strains of bacteria means that this practice should be discouraged.

11.1.4 Preventing scours

To ensure healthy and disease-resistant calves, the importance of good colostrum feeding management cannot be overemphasised. Up to 40% of calves do not absorb sufficient antibodies into their bloodstream within the first 12–24 hr of life because of inadequate attention given to their colostrum feeding (Moran 2002). Such calves are more likely to succumb to infectious scours. Chapter 5 discusses other aspects of colostrum feeding management, all of which can influence calves' susceptibility to scours. In certain countries, vaccines are available to improve the colostrum quality for certain scouring organisms, such as *E. coli*, *Salmonella* and rotavirus. As the demand increases for these vaccines, so will their availability.

Prevention of scours centres around good hygiene and minimising stress. Measures that can be taken include:

- Avoid buying calves from sale yards, because these could introduce disease agents.
- Only buy calves directly from farms which practise good colostrum feeding management and good hygiene.
- Rest transported calves before their first feed of milk.
- Consider vaccinating cows for *E. coli*, *Salmonella* or rotavirus prior to calving, if the vaccines are available.
- Quarantine purchased calves for the first week or so, then disinfect the quarantine area after use, prior to introducing another batch of calves.
- Ensure that calves are protected from extremes of climate, preferably in a shed.
- Carefully plan shed designs to avoid draughts and overcrowding.
- Minimise stresses associated with routine management practices, such as dehorning and castration.
- Maintain strict hygiene by cleaning and sterilising feeding utensils and facilities during milk rearing.
- Develop a routine milk feeding program, with as few people involved as possible.
- Develop an early weaning system to minimise the period of milk feeding.
- Quickly respond to early symptoms of scours, isolate sick calves and rectify the cause.
- Minimise the use of antibiotics, and then only under veterinary supervision.
- Keep records of treatment of sick calves to assist in veterinary diagnoses and for withholding periods if the calf is subsequently culled.

11.2 Pneumonia and other respiratory diseases

Pneumonia is a problem with housed calves, particularly when stocking density is high and ventilation is poor. In the US, it accounts for 15% of the calf deaths from birth to

6 months of age. The shed temperature and relative humidity are the two most important factors influencing its occurrence. Respiratory diseases are more common in cool, damp sheds, although they can also be a problem in hot, dry shed conditions. Typical signs of pneumonia include lethargy, discharge from the nose and eyes, rapid breathing, and a rise in body temperature and pulse rate. Coughing is especially noticeable after exertion because of lung damage and affected calves are more susceptible to further outbreaks and secondary infections.

The control of pneumonia is mainly through improved housing. Poor ventilation leads to condensation, which results in humid conditions and an increase in the survival and spread of infection through water droplets in the air. Draughts of cold air at animal height in pens will aggravate the condition. Regular use of hoses in cleaning pens and laneways can introduce water vapour and blast infectious particles from the manure into the air. High dust and ammonia levels (the latter from urine in poorly drained pens) can cause irritations in the lungs, making these calves more prone to pneumonia.

Early recognition and treatment of affected calves with antibiotics will minimise losses through deaths and poor calf growth. Sheds should be adequately ventilated but draught-free. The use of solid walls to at least 2 m high and then shutters or blinds to control air movement (particularly during cool weather) is ideal. In poorly ventilated sheds, well-positioned exhaust fans can improve air flow without causing draughts. Shed design is discussed in more detail in Chapter 7.

There are other influenza-type, respiratory diseases normally associated with high stocking densities in poorly ventilated sheds and during lengthy sea transport. These are often called 'crowding diseases' in Europe for obvious reasons.

One such disease is infectious bovine rhinotracheitis (IBR). This is caused by a virus and leads to loss of appetite, fever and discharges from the nose and eyes. The muzzle is often bright red (hence the name 'red nose' in Europe) and affected calves breathe with great difficulty. Like all respiratory diseases, secondary infections can confuse their initial cause and veterinary assistance is strongly advisable to ensure the correct treatment.

Pneumonia can also occur in grazing calves and lungworms can play a significant role in damaging the lungs. Adult worms lay eggs in the lung and these are coughed up, swallowed and then passed out onto the pasture. Larvae survive best in cool, wet conditions, so numbers build up on pasture in winter and early spring. Mature cattle have a strong immunity to lungworms whereas calves are very susceptible. Most drenches for roundworms also control lungworms.

11.3 Pulpy kidney and other Clostridial diseases

Pulpy kidney can occur when calves are first introduced to high concentrate diets. It is caused by one of the *Clostridia* bacteria that produces a toxin in the gut, eventually killing the calf (hence the name enterotoxaemia). As with all Clostridial diseases, the bacteria are a normal part of the environment and are impossible to eradicate. The classical sign of pulpy kidney is that the fattest calves (the best milk drinkers) die suddenly and their carcasses rot very quickly. Routine vaccination programs of 'five-in one' vaccines can prevent the disease.

The other Clostridial diseases controlled by 'five-in-one' vaccines are blackleg, black disease, malignant oedema and tetanus. The dilemma with these diseases is that once you have vaccinated, it is hard to prove that it has been worth it – you do not know if you would have lost calves if you had not vaccinated. However, the vaccine is cheap and the cost of a vaccination program is negligible compared with the potential losses incurred through *Clostridia*. It is important to follow the manufacturer's instructions, with regards ages of initial and booster vaccinations.

Calves that have drunk sufficient colostrum from vaccinated cows soon after birth can be partially protected against *Clostridia* up till 6 weeks of age, after which a vaccination at 6–12 weeks of age with a follow up one at least 6 weeks later, gives good immunity. A booster vaccination 12 months later should reduce the incidence of Clostridial diseases in adult cattle and this should be repeated every 3–4 yr. Deaths from *Clostridia* have occasionally been observed following complete vaccination programs, which means that immunity is not always complete.

11.4 Internal parasites and their control

Roundworms and liver fluke are the two most important internal parasites that require attention. Roundworms cause gastroenteritis in young weaned calves. The intestinal worms damage the gut lining, decrease appetite and interfere with the efficient absorption of nutrients. The signs are scours, weight loss, bottle jaw, dehydration and sometimes death. Mild signs include ill-thrift and a dirty tail.

Calves pick up infective larvae while grazing and these mature in the gut within 2–3 weeks, mate and start to lay eggs. There are seasonal peaks of worm burdens, which should be considered when planning drenching programs. Mature cattle are relatively resistant to roundworms, while young stock are the most susceptible.

Worm control depends, firstly, on strategic drenching to suppress worm burdens and to prevent contamination of pasture by worm eggs and, secondly, on integrating drenching with grazing management. Drenching only kills the worms in the calf and does not prevent reinfection. Housed calves have yet to pick up worms and hence do not need drenching.

Drenching programs vary with the area and local recommendations should be followed. Worm test kits are now commercially available in many countries and these can assist with parasite control programs, particularly in determining which drench will be the most cost-effective.

Liver fluke depend on a freshwater snail to complete its life cycle. Acute fluke disease results from massive damage to the liver caused by the immature flukes and can kill calves. Chronic fluke disease is due to the adult fluke blocking the bile ducts of the liver and can lead to weight loss, anaemia, bottle jaw and scours. Adult cattle build up resistance to flukes. If control of snails or control by grazing management is not possible, for instance in irrigation areas, drenches can be used to remove adult and immature flukes before snails become active in the warmer weather. As with all drenches, it is important to follow the labels for dose rates, warnings and withholding periods.

There are other types of worms or internal parasites likely to infest calves in different areas and local advice on drenching and other control measures should be sought.

11.5 Johne's disease

Johne's disease (or paratuberculosis) is an incurable bacterial infection of the intestines. By the time clinical symptoms develop, the wall of the intestine has become thickened and this interferes with the absorption of nutrients from the digested feed. Cows with Johne's disease show progressive chronic diarrhoea and weight loss, ending in death. However, they generally remain bright and alert and maintain a good appetite up to the time of death.

Most infected cows will not show any signs of the disease and stress is important in determining whether the symptoms appear in infected cows. Stresses may include calving, cold weather and feed shortages or moving cows to a different herd or farm. Once Johne's disease is detected in a herd, it is usually well established and there are likely to be other infected carrier cows.

Apparent freedom from clinical cases, even for years, provides no assurance that the herd is free of the disease. Infection occurs during calfhood, but the symptoms are not usually seen until infected cattle are 4–5 yr old. Cattle become more resistant to infection by about 12 months of age.

The disease is spread by a susceptible calf consuming feed, water or milk contaminated by manure from an infected cow. Occasionally, an unborn calf can even acquire an infection from a diseased cow.

Control of Johne's disease depends on separating calves from their dam within 12 hr of birth and then rearing them until 12 months with no contact with faeces from adult cattle. Attention should be given to paddock drainage and not applying reuse water when irrigating calf or heifer paddocks. Drinking water should be supplied from clean sources via troughs, not dams or irrigation drainage channels.

The infective bacteria can survive for up to 12 months in cool, moist conditions. They are destroyed by sunlight and dry conditions.

Johne's disease is a notifiable disease in many countries. In these countries, all cattle showing clinical signs must be reported to local animal health or veterinary advisers. Any herd with a history of the disease can enter a control program supervised by government veterinarians. Infected or reactor cattle should only be sold for slaughter. The Johne's disease status of the herd can affect the ability of cattle to move between zones, states and countries.

Preventing Johne's disease depends on two important factors: firstly, stopping the spread within a herd by good calf-rearing practices, and, secondly, stopping the spread of infection by sourcing replacement heifers from low-risk herds. Calf-rearing systems should prevent calves ingesting feed or water contaminated with manure, while a closed herd is the most effective method of avoiding the introduction of the disease to a herd. Calf-rearing practices should endeavour to:

- calve cows in a clean paddock, because calving pads or heavily stocked calving paddocks present a very high risk of spreading the disease
- separate the calf from the dam within 12 hr of birth
- ensure no cow manure or dairy effluent comes in contact with calves or the calf-rearing area

- feed calf milk replacer or milk from low-risk cattle
- only supply tank, town, or bore water to calves up to 12 months of age – avoid stock dams, troughs or discharge/reuse irrigation channels
- prevent any adult stock entering the designated calf-rearing area – this includes bulls, dry cows, milkers, sick cows, steers, goats or camelids (alpacas and llamas)
- maintain strict hygiene when entering the calf-rearing area – do not introduce dung on boots, clothing or farm machinery such as tractors and bikes
- fence off the calf-rearing area from laneways and milk tanker tracks
- maintain accurate records of calves reared or purchased.

After the calves are weaned and up to 12 months of age, the risk of them becoming infected will be minimised if:

- weaned calves only graze paddocks that have had no adult cattle on them for at least 12 months
- this grazing area is free of any drainage, effluent or sprayed recycled effluent and discharge/reuse irrigation channels are fenced off
- stockyards that are used by adult stock are not used by the calves
- calves sent on agistment are only mixed with stock that have had the same high Johne's disease rearing standards and only graze on areas free from potential contamination.

11.6 Other diseases in calves

11.6.1 Bloat or tympany

Bloat is an over-distension of the abomasum or rumen due to the gas produced by normal fermentation of feed being unable to escape. It can occur in the abomasum of calves fed milk and the rumen of calves fed milk, concentrates or pasture. It is more likely to occur where calves do not suckle the milk and where too much milk is fed too quickly.

The feeding of chopped straw seems to overcome these problems, except where the bloat is caused by an obstruction in the throat or oesophagus. Animals can often show signs of bloat following feeding, but the gas will escape and the rumen or abomasum will eventually return to normal size before the next feeding. If this does not occur, then the use of a stomach tube or a trochar to relieve pressure in the rumen is recommended and veterinary advice should be sought. Prompt action is essential because affected calves can die within an hour after feeding.

Abomasal-induced milk bloat occurs when partially digested milk from a previous feed is enveloped in a clot in the abomasum together with newly drunk milk. Any gases being produced from this partially digested milk cannot escape, causing distension of the abomasum. Rumen bloat can occur in the calf fed milk or milk replacer through failure of the oesophageal groove to close properly or due to back flow into the rumen from the abomasum. This is particularly associated with diets containing certain non-milk proteins, which are rapidly fermented in the rumen.

The author dosing a heifer with a Rumensin capsule.

Abomasal-induced milk bloat appears to be more prevalent with certain types of milk replacers, particularly those incorporating tallow as an energy source. One experienced calf rearer in Australia includes bloat reducing chemicals, such as teric, with the milk replacer at every feeding.

Bloat in grazing, weaned calves is the result of a stable foam developing in the rumen, which traps the bubbles of gas produced by the rumen microbes. The foaming agent is present in the leaves of certain legumes, such as clover and lucerne. Treatment is urgent and affected animals can be drenched with 150–200 mL of vegetable or mineral oil or even butter, lard or cream. If bloat is so severe that the animal cannot swallow, the oil can be inserted directly into the rumen, on the left side, using a wide bore needle. Leave the needle in place to allow some gas to escape. A sharp knife or trochar should be used as a last resort. A small vertical stab wound (2–3 cm long) can be effective and will heal faster than a large hole. Oils, detergents and Monensin in anti-bloat capsules can also be used to control pasture-induced bloat.

11.6.2 Feed toxicities

These can occur through human errors in preparing feeds, supplying inappropriate feeds for calves or providing access to poisonous items in rearing sheds or at pasture. In one instance, Heliotrope poisoning was diagnosed in early weaned calves through contaminated grain being used in the concentrate pellet. In another instance, a producer suffered calf losses through incorrect levels of antibacterial drugs being incorporated into commercial feed preparations.

Calves are very susceptible to lead poisoning and this has occurred through animals licking or chewing painted woodwork and metalwork, discarded paint tins, batteries and painted tarpaulins.

Gossypol, which naturally occurs in cottonseed, can poison calves. It is usually destroyed by the heat process during extraction of the cottonseed oil. Mature cattle are not affected by gossypol (hence can be fed whole cottonseeds) because it is broken down in the rumen. Young calves cannot tolerate it. Cottonseed meal can be high in gossypol, and, for safety sake, should not constitute more than 20% of calf grower rations.

The incorporation of soya flour in milk replacers can create problems if it has not been heat treated to remove the trypsin inhibitor. Trypsin is involved in digestion of milk in the abomasum. Soya flour (like other non-milk protein sources) cannot be used by calves under 3–4 weeks of age.

Vitamin A deficiency has been diagnosed in calves that were rapidly growing and were not early weaned off milk replacer. These calves had low body reserves of vitamin A, were reluctant to eat concentrate pellets and were only offered limited amounts of milk replacer.

Pasture toxins can be a problem in certain regions. Ryegrass staggers can occur in southern Australia, although this rarely causes deaths. Bracken fern poisoning can kill calves through damaging the blood forming tissue in bone marrow. Calves pass blood from their rectum, nose and mouth and respond very poorly to treatment. Deaths have been reported for up to 6 months after calves have been removed from the area.

11.6.3 Grain poisoning or acidosis

This is the result of rapid fermentation of cereal grains and other high-starch feeds in the rumen, leading to excess levels of lactic acid being produced. Affected calves become dull and refuse food, their movements are unsteady, they often scour and bloat may occur. Mild acidosis can be treated by drenching calves with a sodium bicarbonate solution and feeding a roughage-based diet, but severe cases require veterinary attention because death can be sudden. The routine use of sodium bicarbonate and other rumen buffers when feeding high levels of grain should maintain normal levels of acidity in the rumen. The feeding of chopped straw will stimulate saliva production, which buffers the rumen against rapid changes in acidity.

11.6.4 Navel-ill and joint-ill

This is caused by bacteria infecting the umbilical cord soon after birth, particularly where the calving area is heavily contaminated. Unless treated promptly in young calves, it can lead to severe inflammation or arthritis of the joints. Animals with joint-ill are reluctant to walk and stand for only brief periods. Because the infection is carried in the bloodstream to all parts of the body, reduced appetite, diarrhoea and pneumonia may also occur. Navels in all newborn calves should be swabbed or sprayed with diluted iodine (7%) as a precautionary measure and calving facilities should be kept clean. This and other navel abnormalities (such as umbilical hernias) should be apparent when selecting calves for purchase; these animals should be rejected.

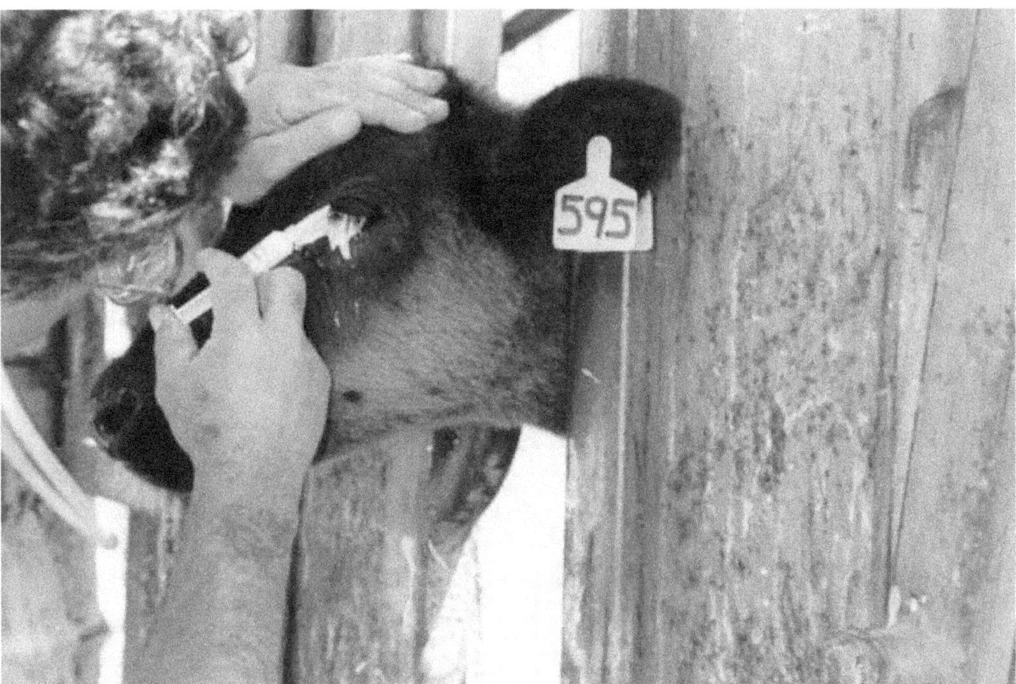

Treating a heifer for pink eye.

11.6.5 Pink eye

Pink eye (or *Moraxella*) is a bacterial infection of the eye, which occurs mostly in the warmer months, possibly as a result of increased fly activity and dust and irritation from young grass. Calves and young stock are more commonly affected than older cattle. The first sign is a discharge from the eye, then it becomes reddened, a shallow ulcer develops and finally the eyeball looks white. Most affected animals recover, leaving small scars that do not appear to interfere with sight. However, in some cases, the eyeball can rupture and blindness results.

Treatment with various ointments, powders or sprays has little effect unless the level of antibiotic is maintained at a high level by frequent applications. Severe cases should be protected by gluing a patch over the affected eye or suturing the eye shut. Control is difficult because little can be done to avoid exposure to ultraviolet light and dust. Fly control helps, but isolation of clinical cases is not effective because the causative bacteria are carried by non-vaccinated cattle.

11.6.6 External parasites

Flies breed in manure and moist feed waste, so these should be removed regularly. Biting flies can cause worry among calves so standard fly control measures (such as fly baits or routine spraying) may be necessary. Lice control using ectoparasite dips or sprays may also be occasionally required due to severe lice infestations, particularly in poor calves.

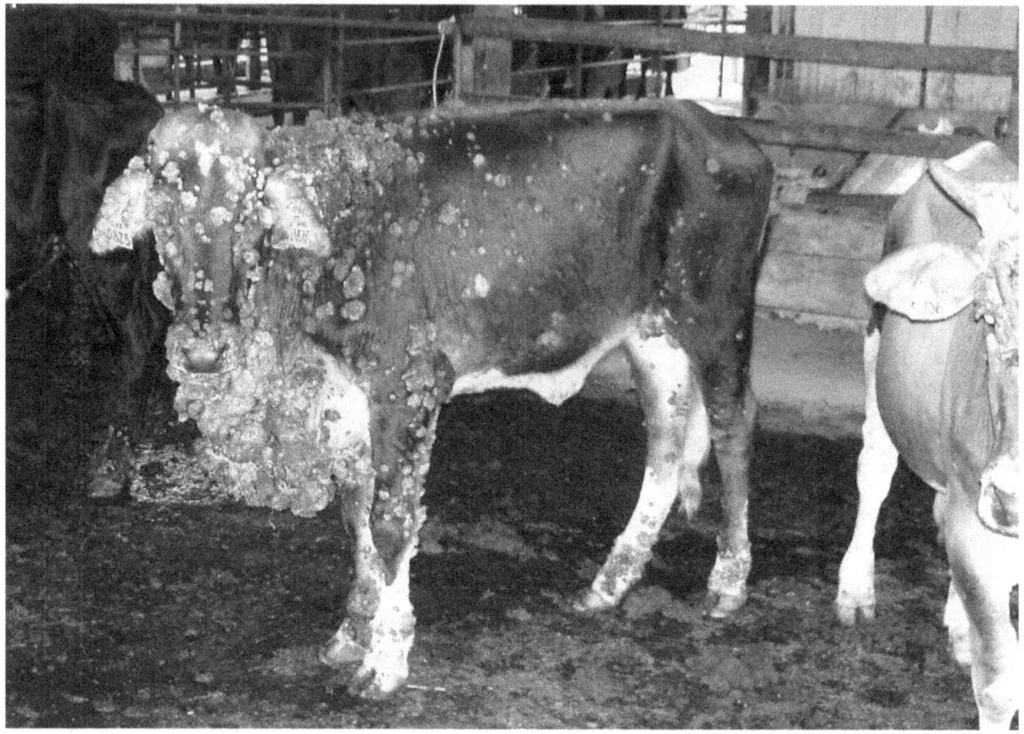

A dairy heifer suffering from a severe skin infestation of papilloma virus.

Ringworm can occur and should be treated with anti-fungal preparations. Severe skin infections such as papilloma virus may not be easily treated and infected stock may need to be culled. Some external parasite treatments should not be used simultaneously with worm drenches, while others should not be used on young calves. It is important to carefully follow the instructions during use.

Cattle tick and buffalo flies can also be a problem with weaned calves. These parasites suck blood and cause skin irritation and also can carry potentially fatal diseases such as tick fever. Therefore it is important to spray or dip calves routinely during the tick season and also to implement recommended grazing management procedures, such as pasture spelling and rotation.

11.6.7 Leptospirosis

This is a bacterial disease that can occur in all farm animals. The 'lepto' bacteria gain entry through the skin or membranes lining the nose, eyes or mouth, or by ingestion. They multiply in the liver, enter the bloodstream, and settle in the kidneys. They are then passed out in the urine. The two most common types of leptospirosis bacteria that affect cattle are *Leptospira pomona,* which can cause abortion, mastitis and milk production losses in mature cows, while it can cause redwater (blood in the urine), jaundice, anaemia and death in calves, and *L. hardjo,* which seldom causes disease in cattle but it does affect humans who catch it from the urine of cattle. Humans show flu-like symptoms including fever, chills, headache, muscular aches and vomiting.

The best way to reduce the risk of milkers acquiring lepto infection is to vaccinate all heifer calves and cows. This will stop cows shedding the bacteria in their urine in the dairy. Calves should be vaccinated twice, about a month apart, once they reach 6 months of age. Combined lepto and clostridial vaccines, called 'seven in one', are the usual form of vaccination. A booster vaccination should be given 12 months later and followed up by an annual booster at the time of drying off. In-calf heifers are a high-risk group and should always be vaccinated.

Depending on the area, other vaccinations or preventative measures are advisable with young calves, such as for *Clostridia* or for tick fever.

11.6.8 Lameness

Lameness is always a potential problem in housed stock. Foot problems can be caused by the environment (continuously wet floors, uneven or broken concrete), poor sanitation, infectious organisms or nutritional imbalances. Leg problems can be the result of traumatic injuries due to poor pen design and overcrowding. Providing stock with soft bedding, such as dirt lounging areas or rubber mats will improve stock comfort, as well as reduce foot and leg problems. Locomotion scores (Moran 2005) can be used to select stock for hoof examination before they become clinically lame.

11.7 Disbudding and dehorning calves

Unlike some breeds of beef calves, all dairy calves will eventually grow horns. It is routine practice to remove these horns or horn buds in early life. Disbudding refers to removing the horn buds in calves less than 2 months of age, whereas dehorning refers to removing the horns and horn buds in older calves, usually older than 2 months of age. There are several methods available, all of which are successful, depending on the experience and skill of the operator rather than the method applied. Calves should be carefully restrained so as not to injure the calf or the operator. Details of these procedures are described below.

11.7.1 Using caustic potash

This should be undertaken within the first week of life. Any contact of the caustic potash with human skin should be avoided and suckling cows should also be kept away from treated calves. The hair should be clipped around an area 2.5 cm diameter over the rudimentary horns. Heavy grease should be spread around the outer edge of the clipped area to prevent the caustic potash running into the calf's eyes. Rub the potash on the skin area immediately over the horn until the hair is removed and the skin becomes red (that is, for about 15 seconds), leave the calf alone for at least a day and don't let it go outside when the weather is rainy.

11.7.2 Using flexile collodion solution

This should be applied within 2 days after birth. Clip the hair around the rudimentary horns and clean it with methylated spirits to remove the fatty covering. Apply the collodion, firstly with a small brush to rub it in, then secondly without rubbing it in. After 48 hr, check whether the collodion film is still present; if not, repeat the procedure.

11.7.3 Using a hot iron (electrical or charcoal)

This should be done 2–3 weeks after birth and is a very good, if somewhat cruel, method. Clip the hair around the horn and place the calf in a head crush to prevent any head movement. Place the heated iron on the horn bud at short intervals, for 5–10 seconds each time. Continue heating until the colour of the tissue around the horn bud turns to deep copper. The heat will kill the growth cells in the horn.

11.7.4 Using a rubber ring (elastrator)

This method is used when the horns have already developed and it is too late to apply other methods.

11.7.5 Using a scoop dehorner

If horns are apparent in young weaned calves and must be removed, they can be using a scoop dehorner. This is very effective and somewhat cruel, and inexperienced operators should seek advice from veterinarians or other animal health specialists.

11.8 Calf management and disease

As far as animal health is concerned, prevention is better than cure. A generous application of drugs is not the right way to deal with health problems in calves; animal health starts with implementing strict hygiene during and after birth. Disease problems with calves seem to be worse during winter and are more frequent in calves with low birth weights. Calves become more resistant to diseases as they get older.

Scouring is more of a problem in milk-fed calves and in group penned rather than individually penned calves and also seems more common in Jerseys than in Friesian calves. Scours are more prevalent in calves fed milk replacer than whole milk, but this could be related to aspects of mixing the replacer or its more variable quality, rather than any more healthy attributes of whole milk. Feeding the youngest, more susceptible calves first each time will minimise any disease transfer from older animals.

Pneumonia, on the other hand, is more prevalent in older calves (6–8 weeks of age) and is not affected by group or individual penning. Early weaned calves seem more susceptible to pneumonia than those fed milk for a longer period. Purebred Friesians also seem more affected than Hereford × Friesian calves. Scouring and pneumonia are more of a problem in calves purchased from calf auctions than those home bred or purchased directly off farm. Sex and weight for age have little influence on the incidence of these diseases. Unfortunately, little is known about the relative susceptibility of indigenous tropical dairy breeds compared with Friesians and Jerseys.

Diseases are more likely to occur in calves subjected to stress than if adequate attention is given to their physical and nutritional needs. Examples of stress include lengthy transport from calf saleyards in overcrowded, unprotected trailers or weaning off milk before the rumen is fully adapted to dry feeds. If there is a smell of ammonia in the rearing unit, better ventilation and/or floor drainage is required to reduce the likelihood of pneumonia outbreaks.

The protection from the passive immunity passed onto the calf peaks 1–2 days after effective colostrum transfer but then it declines. By 2 weeks of age, it has declined enough to increase the calf's susceptibility to bacteria, viruses and other pathogens, before the calf's own immunity increases to an effective level. Therefore the calf can be quite vulnerable to pathogen invasions coming from dirty feeding equipment or other sources between 14 and 21 days of age.

If there is a sudden increase in the number of calves with scours or pneumonia, as well as seeking professional veterinary assistance, it is worthwhile carefully reviewing any possible changes in the rearing management. It may be related to a sudden change in the weather or feeding a different batch of CMR. Perhaps a different member of the farm staff is now feeding the calves (one with a more aggressive personality or a different concept of cleanliness), or some new stock have been purchased from another farm. Such changes in the degree of animal stress or exposure to pathogens need to be considered to ensure that, as well as treating the current disease outbreak, measures are taken to prevent new ones.

The best indicator of health is body temperature. Normal calf temperatures are 39°C in the morning and 39.2°C in the evening. When the body temperature rises, close examination, and often treatment, is necessary. Body temperatures are easily taken with a thermometer in the calf's rectum for 1 min; 20 cm of string attaching the thermometer to a paper clip (which can be clamped onto the calf's hair) should prevent breakage. Electronic thermometers give a meaningful reading within 5 seconds.

Being closely managed, most calves respond whenever people enter rearing sheds. An early sign of disease is general disinterest, in that calves appear listless, apathetic, lack vigour and will not move when approached. Calves standing with the head down and ears lowered are likely to be showing early symptoms of disease. If calves do not stretch when standing after a lengthy rest, they should be carefully observed for other signs of ill health. Loss of appetite, dry and dull coat, sunken eyes, runny eyes and/or nose, fever and difficulty in breathing are obvious signs of ill heath.

Occurrences of disease and deaths are generally lower on farms where owners, rather than employees, rear the calves. Furthermore, deaths are lower on farms where the farm wife rather than the farm husband cares for the calves; death rates on farms where the children rear the calves are intermediate between those of wife and husband. Correct calf rearing – one of the most arduous tasks on dairy or beef farms – requires a genuine concern for the welfare of each animal (in other words, a sense of caring) and a quick recognition of the early symptoms of the diseases described above. One common attribute of all successful calf rearers is TLC. Young calves give out many signals indicating their health and general wellbeing and recognition of these signals becomes second nature to calf rearers with TLC. These are discussed in Chapter 12.

11.8.1 Vaccinate and rest easy

Disease prevention is an investment that can return significant dividends. The corner stones of any disease control are vaccinating against commonly occurring diseases together with good management practices. Deaths and illnesses from commonly occurring diseases, which can be prevented by vaccination, can and do occur in unvaccinated herds or in herds where vaccination is not a routine procedure.

The cost of the loss of a single heifer to one of the Clostridial diseases will far outweigh the cost of the vaccine. Likewise, the cost of a human case of leptospirosis will also far outweigh the cost of annual vaccination of the entire herd.

Vaccination programs should be designed to protect against diseases that occur commonly in the district, plus any specific disease occurring on individual farms. The timing of the vaccination and the selection of the product are important considerations. There are multiple brand names, combinations of products and varying vaccination schedules. It is best to consult your veterinarian or animal health officer for specific vaccination protocols for your herd.

Vaccinations do not provide 100% protection to 100% of the animals vaccinated, but they do increase the level of herd immunity and the level of disease resistance in individual animals. They are not a substitute for otherwise poor stock management practices.

Vaccination can be used in the face of an outbreak of disease. However, the best disease control programs are those that prevent the appearance of the disease in the first instance.

A farmer who has dealt with an outbreak of *Salmonella* and is vaccinating in the face of an outbreak, is very likely to vaccinate the herd with an effective vaccine when he dries off the milking cows, to guard against a similar occurrence next season.

Most of the commonly used cattle vaccines require two initial doses: one as a primer dose, and a second about 4–6 weeks later and an annual booster. Little protection is provided by some vaccines until after the second dose is given. It is best to vaccinate according to the schedule advised by the product chosen.

Labels carry important information about expiry dates, dose rates, injection sites, recommended vaccination programs, storage and occupational health and safety. Use a refrigerator to store vaccines, particularly when the container has been opened. Many vaccines are packaged in multi-dose containers for use with automatic syringes, which must be calibrated to deliver the right dose and sterilised before use. Read the instructions carefully and follow them.

Strict sanitation can help reduce the risk of infection from poor vaccination techniques. Do not inject into areas of the animal hide that are contaminated with manure or dirt. Change needles every five to 10 injections, or any time that a needle becomes blunt, burred or has evidence of blood. Do not put a dirty needle back into the medication bottle. Avoid vaccinating stock during times of stress, including extreme heat. Avoid multiple (greater than two injections) vaccinations when possible.

Cattle diseases for which vaccines are available in most Western dairy industries include:

- *Clostridia*
- leptospirosis
- *Salmonella*
- *E. coli*
- vibriosis (or campylobacteriosis)
- pink eye (or *Moraxella*)
- pestivirus
- Mannhemia (a respiratory disease)

- anthrax
- foot and mouth disease
- Johne's disease.

11.9 How to recognise sick calves

Before rearers can recognise sick calves, they must know how healthy calves behave. This allows them to be on the alert for subtle changes in calf behaviour before clinical signs of disease become obvious. They should never be complacent about changes in calf wellbeing and behaviour.

Calves charging your knees and running around the pen are healthy. Such calves rest in a curled-up position with feet tucked under and heads back along the body. They appear relaxed with regular breathing. Some healthy calves may also rest flat on their sides.

11.9.1 Signs to look for

Each day look quickly over each pen of calves, then be more specific and check suspect calves' noses for dampness and ears for temperature. Sick calves often have dry noses and higher than normal body temperatures and a depressed attitude. Listen to their breathing, noting any 'rattles' or laboured breathing. Lift their tail and note the state of any faecal residues. Look at their feet and legs. For the first week to 10 days of age, check the navel area for signs of inflammation and swelling. This inspection should be undertaken as part of your daily routine.

Calves resting in the corner of pens, with their head turned away from pen mates should not be ignored. Get the calf up. If it stretches, it is okay. If it does not, it may require further attention. Sick calves show general disinterest, become listless and apathetic, lack vigour and often do not move when approached. They may stand with their ears lowered and head down. Chapter 12 lists many of the changes in behaviour or physical appearance that allow each calf to 'communicate' how it is feeling.

In summary, watch out for any calf:

- that is slow to get up and eat
- that does not immediately begin eating vigorously
- that lingers over the milk bucket longer than others
- with coloured or opaque nasal discharge or with loose manure
- that is coughing
- that is straining to urinate.

Calves must be kept in a stress-free environment. It is difficult to identify changes in the behaviour if calves are kept in conditions where they look miserable and hunched up because of cold stress.

11.9.2 Keep records to help identify problems

Records should be kept of changes in the intake of milk and concentrates and of fluctuations in growth rates. Body temperatures should be recorded in suspect calves to assist with disease diagnosis.

The reasons for outbreaks of scouring must be tracked down. It may simply be due to a change in feeding or management routine, in which case little further treatment is necessary. However, if scours persist, veterinary diagnosis should be sought. Only use antibiotics under instructions from your veterinarian.

11.10 What should you do with sick calves?

It is up to the calf rearer to decide whether to do nothing, to treat the animal themselves or to contact the veterinarian. The other essential step to take if the disease is infectious is to stop the spread the disease by moving the sick animal to the isolation or 'hospital' pen. For all diseases, diagnosis should be obtained and post-mortem reports should be sought for any unexplained calf deaths. Veterinarians can send samples to local veterinary laboratories for pathological examinations.

Like any professional, most veterinarians have specialised interests in their disciplines. When attending dairy farms in the US, veterinarians spend less than 5% of their time working with pre-weaned calves. It is not uncommon in intensive dairy regions, for farmers with calf-rearing problems to seek assistance from a particular clinic and a particular veterinarian because they have been happy with the way they have previously worked together. When rearers employ a veterinarian, they should not be afraid to follow him or her around the shed asking questions about the techniques used for diagnosis. When clearly explained, most of them are obvious in solving the disease problem encountered, such that in the future, rearers could undertake most of the diagnosis themselves.

When commencing a new calf-rearing operation, or reassessing an existing one, it is always a good idea, if possible, to find a veterinarian that you can work with. It is important to develop a disease prevention and management program, before any problems arise. Good calf rearers should rarely need a veterinarian. They should be able to identify the early signs of ill health then act on them before the calf requires much treatment. In my experience, the best calf rearers are women, because of their more empathetic and caring natures, and the best women calf rearers are hospital nurses, because of their training to anticipate health problems before they occur.

11.10.1 Assisting the veterinarian

There is much that farmers can do to assist veterinarians or other animal health professionals in diagnosing and treating diseases. These include:

- Examine the history – for how long has the calf been sick, how rapidly did the sickness develop, have new stock recently been introduced to the farm, have there been any recent environmental changes (such as hot weather)?
- Examine the animal – use the checklist above.
- Examine the environment – are other animals sick, could there be poisons or mineral deficiencies involved, what is the water quantity and quality, what are the physical conditions, such as lanes or yards, that cause traumatic problems?
- Clearly identify the sick animal – either by recording the ear tag or placing some identification on the calf, such as coloured string around its neck or spraying paint on

its head. This will serve two purposes, firstly ensuring the veterinarian attends to the correct animal and assisting with any follow up treatment.
- Prepare for the veterinary visit. Record the important major symptoms, have the sick animal easily accessible with suitable restraining equipment available, listen carefully and take notes for any follow up treatment.
- Once the disease has been diagnosed and its cause, symptoms and treatment identified, it is important to develop a control program to reduce its incidence in the future. Veterinarians can assist with such a control program. For the farmer, prevention is just as, and often more, important than cure.

Once the veterinarian has provided treatment, it is important to continue to nurse sick calves. Supportive care that can be provided includes:

- Provide good nutrition to treated stock, with freely available water and quality palatable forages and concentrates.
- Ensure access to water and shade and clean bedding.
- Maintain a low stress environment, with adequate housing, removal of competition from other stock, remove parasites, provide pain relief, and clean and dress any wounds.
- Take special care of stock unable to stand, provide support such as small hay bales to stop them rolling onto their side, move them to dry, warm shelter with good footing, in case they try to stand.

11.10.2 Their long-term future

Unhealthy calves are likely to grow into unhealthy cows – and unhealthy cows cost money. They have higher drug, veterinary and labour costs, and also reduced performance: that is, poorer lifetime milk yield and fewer calves born. Not only do unhealthy cows cost more money through fewer lactations in the herd, their higher culling rates increase the need to rear more replacement heifers.

What then should you do with sick calves? Our inherent nature, it to provide them with all the TLC, required veterinary assistance and drugs until they are up and about running with their pen mates. But what then? Should you keep them and grow them out or sell them at the first opportunity?

How much do sick calves cost? It is relatively easy to record the cash costs of treatment, such as veterinary visits and drugs. It is more difficult to cost out the extra time and care required during treatment and recuperation. For example, US researchers found that each sick calf required, on average, 53 min of extra care before recovery occurs. However, it is the long-term effects on heifer health and subsequent performance that are near impossible to quantify. These are much higher than the costs and labour during treatment.

Overseas studies have confirmed the concern that sick calves have poorer performance as adult cows. For example, in Canada, heifer calves that were treated for scours were two to three times more likely to be sold prior to mating and three times more likely to calve down as 30-month-, rather than 24-month-, old heifers. Furthermore, those that were treated for pneumonia during their first 3 months of rearing were two to three times more likely to die within this 90 day period. Such problems are reflected in wastage rates, which we can describe as the proportion of live heifer calves born that are culled or die before their second calving. Overseas targets are for 20% wastage, whereas surveys in Australia have recorded wastage as high as 35%.

So, when should you decide whether to sell a calf or not? How sick should she be before you have to decide that she is never likely to be a really profitable member of your milking herd? There is no easy answer to this quandary. I suppose all we can conclude is that the more attention a sick calf requires during treatment, the less likely she will make you money as an adult cow.

Some astute farmers are adamant that every sick calf should be disposed of, either by humane slaughter or sale as a cull. The problem with selling such animals is that, unless the animal goes straight to the abattoirs, the problem of potential poor growth is just passed onto the new purchaser. However, one would expect that an astute calf rearer is unlikely to purchase a recovered calf.

How many farmers actually document which calves get sick, the degree of treatment required for their recovery, their age and reason for culling? If this record keeping became routine, farmers would then know how many lactations such animals are likely to remain in their milking herd. This could eventually provide a valuable benchmark for them to make the decision to cull recovered calves or let them join their healthier heifer calf mates. Only then can farmers make truly objective decisions as to the fate of their previously sick heifer calves.

11.10.3 Overuse of antibiotics

Antibiotics should only be used as a last resort with sick calves. Most scours can be successfully treated with a program of electrolyte fluid replacements. Because many of the scour-causing agents are not even bacteria, antibiotics will not kill them. The cautions about overuse of antibiotics were discussed earlier in this chapter in Sections 11.1.2 and 11.1.3, and also in Section 8.6.4.

Whenever antibiotics are used, dairy farmers have an obligation to ensure that they do not get into the human food chain. The most common method of antibiotics being potentially consumed by humans is through raw milk or the meat from slaughtered calves. As well as detailing the most appropriate dose rates for treating sick dairy stock, the labels of all antibiotic containers should include what is called, the withholding period. The withholding period is the minimum number of days following the last antibiotic treatment, before which milk can be sold to milk processors and treated stock can be slaughtered for human consumption. Prior to the withholding period:

- the milking cow can still be secreting antibiotics into the milk
- the calf to be slaughtered has not excreted all the antibiotics from its body.

This then requires farmers to keep good records of stock treatment and adhere to any required withholding periods. Most milk processors can test the raw milk for antibiotics, and if any are found, downgrade or discard the milk, generally at the expense of the farmer or dairy cooperative. Similarly, meat from slaughtered calves can be tested for antibiotic residues. In many Western countries, farmers are penalised if their slaughtered calves test positive for antibiotic residues.

When a need has been determined, antibiotics can play an important role in the treatment plan, but it is irresponsible and ultimately costly to administer antibiotics to all sick calves. Effective use of an antibiotic is also dependent on knowledge of the most appropriate dose, administration method and withholding periods.

Treatments for mastitis in milking cows or scours in calves are the two most common forms of antibiotics entering the human feed chain. Most farmers are aware that mastitis treatment almost always involves the use of antibiotics. Allowing such milk to be mixed with non-antibiotic milk prior to the withholding period is often a conscious decision by the farmer in an attempt to maintain farm income.

However, many farmers are unaware that some commonly used scour treatments also contain antibiotics. In many situations, such drugs are poured into smaller containers without details of the label being similarly transferred. It is the obligation of the supplier of veterinary drugs – usually a veterinarian or someone trained in animal health protocols – to ensure farmers are made fully aware of the drugs they administer together with any associated warnings of what they contain. Furthermore, treated calves should be easily identified and isolated to reduce the likelihood of cross contamination with other non-treated stock. Feeding utensils should also be thoroughly cleaned and sterilised so they do not pass on the antibiotics when used to feed other calves.

11.10.4 Probiotics

This is a term used for preparations of liquid or powder based on bacteria that normally inhabit the stomachs and intestines of healthy stock. Probiotics are being promoted as benefiting sick calves, such as scouring ones, and also as general health 'tonics' for healthy stock. Manufacturers claim probiotics increase the numbers of bacteria in the gut and in so doing, improve the efficiency of feed digestion.

Although theoretically it would seem to benefit calves suffering from infectious scours, by reducing the undigested feed residues used by pathogenic bacteria to grow and multiply, there are few truly independent scientific studies that support their use. There may be some benefit with improving growth and feed efficiency in healthy calves and heifers; however, it is often difficult to justify probiotics without a careful analysis of the costs and returns of their application.

11.11 Maintaining a healthy calf shed

It is important that calf sheds be maintained as disease free as possible. The main avenues for introducing new diseases are via calves purchased off farm, contamination from adult stock on farm (such as Johne's disease) or from service providers visiting the calf shed. Biosecurity is the term used to describe the restricting of access by other livestock and personnel.

One US county has recently formed a group of concerned dairy professionals to categorise service providers into high and medium risk, based on their contact with livestock and farm equipment. This approach could be used for calf-rearing operations.

In that case, high-risk people would include veterinarians and other calf consultants, operators of mobile calf scales, drivers of feed trucks, dead stock removal personnel, visiting farmers (both local and from other areas) and sales representatives and service personnel (with access to other calf operations). Medium risk people would include consultants, sales representatives and service personnel (without access to other calf operations) and non-farm visitors.

The committee suggested the following protocols for high-risk service providers:

- Wash and disinfect boots before and after visiting the calf shed.
- Park vehicles away from the shed.
- Use clean and disposable coveralls at all times.
- Wash and disinfect all equipment.
- Collect and deliver livestock in clean trailers and trucks.
- Pick up dead stock away from the calf shed.

Medium-risk service providers could be asked to follow the first two of these protocols. Other actions that could be taken undertaken include:

- Place signs indicating 'Biosecure area – do not enter without permission' at the calf shed and 'No visitors, sales, services people allowed without an appointment' at the front gate. This would necessitate listing a contact person and telephone number.
- Feed delivery personnel should wear plastic boots and be encouraged to use non-returnable bags.
- Place the bulk feed bins away from the calves, if possible.
- Ensure that all calves purchased off farm are shedded separately to the home calves, at least until they are 2 weeks of age.
- Control the movement of people, animals and equipment onto the farm. Some diseases are spread on clothing and boots. If equipment is borrowed from other farms, it should be cleaned prior to use. Keeping transport or service personnel away from the main herd area, especially the calf shed, or at least providing them with footwear, will also reduce the likelihood of introduced diseases.

While not all of these measures are practical on all farms, they should at least be considered, and some should be implemented to reduce the risks of introducing new diseases into the calf shed.

11.11.1 Responsible drug handling

Responsible drug handling and application are the key to any successful animal health program. Drugs should be sourced from veterinarians or registered agricultural merchants because medicines obtained from other sources may not be safe or effective. They should be stored correctly in accordance with the instructions on the label. Storage temperature is critical for some medicines, especially vaccines. Light can damage others. Make sure they are stored securely and locked where practicable. Keep out of reach of children, animals and anybody not supposed to handle them.

Keep records of:

- how much was purchased and when
- the batch number and expiry date
- when it was used and on what type of stock
- the withholding period (start and end dates) for sale of milk or for slaughter.

Veterinary drugs should be stored in a cupboard away from the sun.

Use the drugs only on animals recommended on the label. Dispose of unused medicines safely when treatment is finished. If using disposable needles and syringes, dispose of them after use in a safe container. For other reusable equipment, clean and sterilise before and after use. If using a syringe that requires filling from a bottle between doses, use one sterile needle left in the bottle during use to fill the syringe and a separate needle to inject the animal. Use a separate needle, or at least sterilise it, between animals. Make the injection site through an area of clean and dry skin.

High standards of sanitation are required at all times to prevent rapid spreads of infectious diseases in both young stock and milking cows. The most effective way of destroying disease-carrying micro-organisms is cleaning and disinfecting (sterilising or sanitising). However, the latter has little effect unless the surface is first cleaned. The best cleaner and disinfectant depends on the type of surface.

11.11.2 Biosecurity when purchasing new stock

Most dairy farmers purchase new stock, so it is important to plan their introduction to minimise the risk that they will introduce infectious diseases. Three factors are important in reducing this risk:

1. Protecting the herd with proper vaccination.
2. The source of purchased stock, including how they are transported to the farm.
3. The method to introduce the new stock to the rest of the herd.

When purchasing new stock, whether they be cows, heifers or newly weaned calves:

1. Ensure their health status is known.
2. Where possible, ensure details of their vaccination program is known.
3. Avoid purchasing stock from unknown sources or stock that have mixed with other cattle before sale, such as from cattle markets.
4. Purchase heifers because they can be more easily quarantined and are less likely to have mastitis.
5. Calves purchased should be kept separate for at least a week with no signs of disease.
6. Transport purchased cattle preferably in the farmer's vehicle, in a clean truck or trailer.

Other steps that can be taken with newly purchased stock include quarantining them in an area separate from other cattle on the farm, using a medicated foot bath before allowing them to enter the herd, and vaccinating them during the quarantine period to make sure they are integrated into the farm vaccination program.

Other biosecurity measures that should be considered include controlling the movement of people, animals and equipment onto the farm. Some diseases are spread on clothing and boots. We all know how much calves love to suck on every part of you they can reach. Clean overalls for all employees, for use only in the calf shed, will reduce any disease spread. Disposable gloves could be used when closely examining any sick calf. If equipment is borrowed from other farms, it should be cleaned prior to use. Keeping transport or service personnel away from the main herd area, especially the calf shed, or at least providing them with footwear, will also reduce the likelihood of introduced diseases. Some disease conscious calf-rearing operations even insist that all employees and visitors walk through a sanitising footbath as they enter the calf shed.

11.11.3 Animal and human health

Government veterinary services must maintain surveillance of infectious or notifiable diseases, such as rinderpest, foot and mouth disease and contagious pneumonia, through vaccination and quarantine measures. However, as farms become more intensively managed, non-infectious diseases play a more important role in limiting cow performance. These could be called disease-causing risk factors such as undernutrition, poor hygiene and other management factors affecting herd productivity.

Farmers are only interested in herd health programs when the link with production is clear, such as declining milk yields, increasing mortality and poor reproduction. Mastitis, for example, is hard to manage because the sub-clinical form is very prevalent, difficult to detect and causes higher milk losses per affected cow. It is often difficult to incorporate economic parameters in such programs because of the small number of stock, and hence the influential impact of the performance of each animal.

There are many diseases that can be transferred from intensively managed livestock to humans. These include *Salmonella* (and other calf scour-causing micro-organisms), ringworm, mange (and other skin diseases) and leptospirosis. Children are particularly susceptible because of their affinity to young calves, and their poor understanding of human hygiene. Another potential hazard for young children are veterinary drugs and chemicals used for cleaning or sanitation. These should be stored in a secure place.

12

Communicating with the calf

This chapter describes how to interpret the wellbeing of a calf from its behaviour and appearance.

The main points in this chapter

- Calves give many signals that indicate that they are in good (or poor) health and quick observations by the rearer can help treat any disease conditions early.
- It is important for rearers to form bonds with their calves so the calves will cooperate more fully, particularly following treatment for diseases.
- It is important for rearers to develop their own 'dictionary of calf language'.
- Rearers should learn to closely observe and interpret changes in both calf appearance and in their normal behaviour, which might be symptomatic of stress.
- Calf scours comes in many forms and colours, all of which can be used to help diagnose a cause.
- It is important to understand how calves react to people so that rearers' management practices can be changed accordingly.
- Farm owners and managers should communicate with their calf rearers.
- Developing a set of standard operating procedures, and writing them down, can help maintain consistency in managing and training new staff in the desired skills of calf rearing.
- Contract calf rearers can provide the right motivation and skills to rear calves better than staff on the home farm.

Success or failure in raising calves depends to a great extent on the rearers' attitude to the calves and their ability to react promptly to the calves' numerous signals (Figure 12.1). Interpreting these signals is a skill that can be easily learnt. Recent developments in calf rearing are directed towards reducing the average time spent with each calf. In many cases, at least part of that time saved would be well spent in observing the calves more closely.

Dairy farmers should develop a critical eye to observe the calves' behaviour and react to any signals the calves are showing. A good farmer understands these signals and needs to observe, analyse, improve and react. As a farmer once told me, 'A good calf rearer knows which calf will get sick tomorrow'.

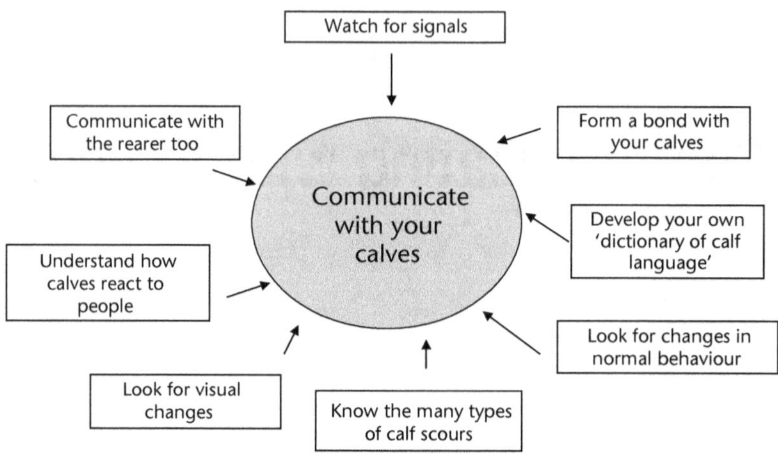

Figure 12.1. How to communicate with your calves

12.1 Signals to watch for

Do not let your senses idle when handling calves because many potential or actual problems may be picked up by close attention. Table 12.1 presents some of these.

Quietly talk, hum or even sing while you work, to help the calves become familiar with your voice. Scratch each calf behind its ear or underneath its throat. Later on, this rapport could help you convince a sick animal to eat or otherwise cooperate with you. Forming a bond with your calves is valuable in the long run.

When moving groups of calves around, remember not to try to move them from behind. Calves can see in most directions, but cannot see directly behind them, so always drive animals from the side. If you are directly behind them, they are likely to stop and turn around to see who is behind them. They always go in the direction they are headed, so calves should face the direction they are to be driven before any pressure is put on them to move.

When trying to get them through a gate, stand beside the gate. Once the calves are looking at you and facing the gate, step towards them and they will run through it to escape from you. Walking with the calves slows them down, whereas walking against the direction in which they are moving speeds them up.

Paying attention to the signals calves are constantly giving you enables you to improve your communication with them. Possibly, in time, you may develop your own 'dictionary of calf language'. Although it may not have too many entries, it may be of value to you from time to time.

There are veterinary treatments for most diseases whose symptoms become apparent through changes in calf behaviour and/or appearance, such as those described below. This chapter simply lists symptoms that may indicate the cause of the stress, while Chapter 11 deals with some of the veterinary treatments.

There are many suggested treatments for problems during calf rearing that do not always appear in the standard veterinary texts, such as using charcoal or cornflour to

Table 12.1. Using your senses to monitor the wellbeing of calves and conditions in the calf shed.

Sense used	Indicators of ill health or wellbeing
Eyesight	Bright and alert eyes
	Droopy or upright ears
	Soft and shiny skin and coat
	Panting and rapid respiration rates
	Abnormal discharges from eyes, mouth or body
	Whether navels and joints are swollen
	State of faeces residues on calves' back legs (colour and consistency)
	State of faeces on floor (runny, too clumpy)
	Any excess feed residues
	Willingness of calves to eat and drink
	Any abnormal calf behaviour
	Proportion of calves resting, standing or moving around
	If calves stretch when they get up
	General state of calf shed (drainage, ventilation)
	General tidiness and cleanliness of calf shed
Hearing	Grinding of teeth
	Bellowing
	Laboured breathing
	Coughing
	Unsettled calves moving around pens
	Dripping taps or water troughs
Smell	Abnormal odour of calves breath
	Odour of faeces
	Any other abnormal calf odours (infected hooves)
	Odour of whole milk or CMR powder
	Odour of bedding
	Odour from mouldy feeds
	Odour of air, hence state of ventilation
	Odour coming from poor drainage
Taste	Taste of whole milk or CMR solution
	Taste of concentrates and forages
Touch	Whether noses are dry
	Whether ears are warm, hot or cold
	General level of heat or cold stress for calves
	Any abnormal draughts
	Whether air is too damp, indicating poor ventilation
	Temperature of milk or CMR solution

reduce the incidence of scouring, or using ginger as a tonic for sick animals. Some of these are based on accepted medical principles, such as reducing the rate of movement of gut contents in scouring calves through increasing its viscosity. Others have evolved over

Calves can communicate with their rearers in many ways.

the years from folk medicine and as yet are either not fully understood or may even be discounted by mainstream veterinary science. That is not to say they do not work.

There is no harm in discussing them with veterinarians and experienced calf rearers. In fact this should be encouraged in the interests of better calf husbandry.

12.2 Changes in normal calf behaviour symptomatic of stress

- **The calf is charging your knees, running around the pen.** These signs characterise a healthy calf. It will do well for you.
- **The calf has a poor appetite at birth.** A disinterest in food shortly after birth is often related to traumatic events preceding or surrounding the birth. Do not wait for the problem to correct itself. Offer the calf high-quality colostrum by stomach tubing two or three times a day. Continue this treatment until the calf is ready to eat on its own. To prevent recurrence, review the nutritional program for the milking herd, particularly during the 60-day, non-lactating period preceding parturition, and correct for any deficiencies in protein, minerals and vitamins.
- **The calf is resting in an abnormal position.** About 1 hr after feeding, walk through the shed and observe the calves. Healthy calves rest in a curled-up position with feet tucked under and heads back along the body. They appear relaxed with regular

breathing rhythms. Any deviation from this standard should be judged with suspicion, although some healthy calves just rest flat on their side. A calf that lies flat on its side may need propping up to prevent the fluid from the stomach draining back to the oesophagus and then into the lungs. If its neck is stretched directly ahead, with front feet tucked squarely under its chest and its shoulders humped quite high, *Salmonella* may be a problem. A calf that curls its back up and down, with the nose pulled close to the body may have a sore throat.

- **The calf does not stretch when standing up after a rest.** Following a lengthy rest, a calf will generally stretch its legs when aroused and get up. If it does not, pay particular attention to it such as ensuring it drinks with its pen-mates. Lack of stretching is often the first sign of ill health.
- **The calf is disinterested in the food and surroundings.** This animal could be telling you that you have betrayed it in the past and you didn't react to its previous signals. The road to recovery would not be easy. Seek a diagnosis from your veterinarian and treat the calf as advised. Try to remove the original source of stress and, following veterinary advice, consider treating the infection (if this is the likely cause) with a broad-spectrum antibiotic if advised. Give the calf electrolytes to prevent dehydration, using stomach tubing if necessary. If a digestive disorder is the cause, a teaspoon of ginger (apparently known as Canadian tonic) may restore the calf's interest. Inject the calf with vitamin B and check that vitamins A, D and E had previously been given.
- **The calf lies with its neck stretched, front feet tucked squarely under its chest and shoulders hunched high.** This calf is likely to be suffering from *Salmonella*. Its temperature could have risen to 41°C and the calf would be generally weak and depressed. Foul-smelling diarrhoea, often green in colour, contains blood and later, pieces of intestinal lining. Unless the disease is identified and treated very early, 60% of infected calves usually die, and those that survive will perform poorly. Do not introduce new calves into a pen until all its previous inhabitants have moved on and it has been disinfected. To prevent infection in humans, high standards of personal hygiene must be maintained. This is difficult on seasonal calving farms where alternative calf-rearing sheds may not be available, but on year-round calving farms, a temporary smaller shed could be constructed.
- **The calf gulps the milk and chokes on it.** This occurs in calves that are underfed, under stress or have to compete for milk. Some calves will plunge their heads into milk buckets, splashing it all over the floor and inhaling some into their lungs. Offer a small quantity of milk at a time and, if possible, separate the calf from others. The gulping and choking usually stops once a regular feeding program is established.
- **The calf stops eating starter pellets.** Partial or complete refusal of calf starter may indicate that energy needs have been fully satisfied by liquid feeds. Stress or a severe case of digestive or respiratory disorders may have the same consequences. Prolonged treatment with certain drugs (such as sulpha drugs), particularly given orally, may impair rumen microbial activity and thus temporarily reduce starter intakes. If all calves refuse it, feed quality may be the problem. Check for mouldy and/or musty concentrate ingredients (if using an on-farm mix). Excess minerals can also reduce its palatability, which can be improved with molasses.

- **The calf is kicking the belly area with its hind legs.** This indicates pain in the abdominal area. The source of the pain could be twisted stomach, constipation, urinary calculi (kidney stones) or bloat. Desperately seeking relief, the calf with a twisted abomasum frequently lies down and jumps up. To help, place the calf on its back on heavy straw bedding and, holding the front and hind legs, roll it from side to side a few times. A constipated calf frequently strains and bellows loudly while trying unsuccessfully to pass manure. If the calf is not drinking, give it water or electrolytes using a stomach tube. Urinary calculi could be suspect if substantial deposits of mineral salts are deposited on the sheath (of bull calves) and the calf tries to urinate frequently.
- **The calf is unable to stand or even raise its head.** Examine the calf thoroughly for possible soreness, such an injured knee, displaced joint, infected navel, and so on. If it cannot even raise its head, this may indicate complete exhaustion due to a long battle with pneumonia or scours. If the calf has a normal body temperature and a history of good health, it could be due to muscular dystrophy through a deficiency of selenium. Once treated with selenium and vitamin E, the calf could be back on its feet within 24 hr.
- **A calf is drinking the urine from other calves.** Pizzle sucking is a vice usually related to an unsatisfied sucking instinct, particularly in early-weaned calves. This problem can be largely overcome by tethering calves during milk feeding, using rubber teats or keeping calves separate till well past weaning. Hanging a piece of chain in pens may also be effective in group housing systems.
- **The calf is resting in the corner of the pen, with its head turned away from its pen mates.** This animal should not be ignored. First, get the calf up and if it stretches, it is fine. If it doesn't, then it requires attention. The calf may be at the bottom of the social hierarchy in the pen and should be moved in with smaller or less-aggressive calves. If the shed offers poor protection against the wind, this corner may be the warmest area of the pen and all calves will tend to congregate there.
- **The calf is shivering with its hair standing up along its back.** This animal is suffering from cold stress and should be better protected from draughts or provided with thick, dry bedding and a source of heat. If only one or two calves show these symptoms, check their body temperatures. Calves may shiver in very cold weather if fed milk at too low a temperature.
- **The calf has an increased breathing rate at normal air temperature.** Increases in respiration rate in hot weather are expected. Some of the best gaining calves can have higher than normal rates as they consume more feed and require extra oxygen for its assimilation into body weight gain. Normal breathing rates in temperate regions are 56/min at 4 days, 50/min at 14 days and 37/min at 35 days of age; such data is yet to become available in tropical areas, but anything less than 50 or 60 breaths/min would be considered normal. In the majority of cases, calves with increased breathing have reduced lung capacity due to respiratory problems such as pneumonia. Increased body temperatures often accompany these disorders.
- **The calf is standing with its front legs spread out and head stretched ahead.** These are important signs of a lengthy bout of pneumonia. Only a portion of its lungs are functional and the spreading of the legs allows the calf to try and secure more volume

for the lungs to make breathing easier. If the above symptoms are accompanied by a heavy discharge from the nose and frothy saliva is running from the mouth, then damage of lung tissue has probably been irreversible. Another symptom of pneumonia may be an arched back, with the calf moaning.

- **The calf grinds its teeth.** You are dealing with a calf which has lost the will to live after suffering from extended pneumonia, scours and/or chronic bloat. The chances that you will save it are very slim. Give the calf an isolated, warm pen with fresh feed and water. Do not spend too much money on additional medication. Another pre-death symptom is temporary or permanent loss of eye muscle control, described as 'sky or star gazing'.
- **An otherwise healthy calf is suddenly found dead.** Though many causes may be involved, lead poisoning is a likely suspect. Consumption of only 150–200 mg of lead represents a lethal dose. Sources of lead include discarded car batteries, certain herbicides, discarded paint tins or painted woodwork. Calves that have ingested only small quantities of lead and are still alive, look dejected, dull and have sunken eyes. They often show abdominal pain and grind their teeth. Treatment is available from veterinarians. Pulpy kidney is another possible cause of very rapid death.

12.3 The many types of calf scours

The characteristics of calf faeces can be a good indication of the type of digestive disorder being suffered. Here are a few examples:

- **Blood is present in the faeces of the newborn calf.** The inner lining of the intestine of a newborn calf consists of immature cells that are replaced within a few days after birth by more permanent ones. During this period, the fragile blood vessels can easily break. When the broken vessel is close to the end of the gut, bright red blood appears in the faeces. If bleeding is excessive, an injection of vitamin K (a blood coagulating agent) can be given. Otherwise the occasional appearance of blood in faeces should not be of great concern. When excessive bleeding is accompanied by high temperature and scours, coccidiosis or salmonellosis may be occurring.
- **The calf has white or yellow scours.** This suggests that a number of the classic pre-scour signs, such as loss of appetite, depressed appearance, facial hair standing on end, were missed. Among the possible causes of scours are inadequate colostrum, overfeeding, overcrowding, poor sanitation and general stress. If the scouring was the result of inferior milk replacer, the volume of faeces will be usually large with a gelatinous consistency.
- **The calf has watery scours.** Mild cases of watery scours, usually lasting 6–12 hr, are often seen in purchased calves after about 5 days in the rearing shed. They are connected with the change in diet, stress or by slight overfeeding. Sometimes the faeces contains blood stains originating from a broken vessel.
- **The calf has bloody scours and it is straining to pass manure.** The presence of blood in the faeces may be of no significance or it may indicate serious infections from *Salmonella* or *Coccidia*. If a calf that is 14 days or older has a normal or slightly

elevated temperature, but its watery faeces contain large clots of fresh blood or dark tarry blood staining, it is likely to be suffering from *Coccidia*. The infectious agent is a common opportunist that is present in more than 50% of healthy calves. The incidence of coccidiosis rises on farms where calves are subjected to early confinement and are exposed to massive infections at an early age. This is a stress-related disease and usually indicates a poor rearing environment.

- **The calf has loose, dark brown stools.** This usually indicates bleeding from lesions and ulcers in the abomasum or a serious infection in the digestive tract. When bleeding takes place in the abomasal area, medication seldom helps. Use gastric and intestinal protectants containing kaolin, pectin or bismuth. If the calf is eating solid feeds, reduce the acidity in the gut by feeding less grain and more roughage.

12.4 Visual changes in calves symptomatic of stress

- **The calf's eyes are bulging.** The eyes in some newborn and young calves protrude from the eye sockets, giving it an appearance similar to people with a thyroid disorder. Fortunately, bulging eyes in calves indicate a good supply of body fluids and a scour-free history. If the calf is healthy and attentive, bulging eyes should not discourage you from purchasing it.
- **The calf has droopy ears.** This animal is likely to be running a high temperature because of pneumonia or a digestive disorder. Check the temperature and, if it is high (above 39.7°C), the calf should be immediately treated with a broad-spectrum antibiotic. If one ear droops, check for external parasites such as lice and treat them with a few drops of hydrogen peroxide. If the base of the ear is swollen or tender, use a teaspoon of warm olive oil and gently work it into the skin. Any discharge from the ear usually indicates an infection that could be treated with penicillin.
- **The calf has facial hair standing on end.** When first observed in a previously healthy calf, this usually indicates an imminent digestive disorder. It is likely that the calf will be scouring within 24 hr. Skipping one milk feed (if twice daily feeding) and replacing it with electrolyte may help. If the calf was purchased with facial hair standing on its end, or if it is a permanent fixture, the calf has possibly had lengthy pneumonia and is still not feeling well.
- **The calf has sunken eyes and its skin has lost its flexibility.** The problem is dehydration and it has not been recognised or treated for the last few days. Prolonged scouring leads to substantial loss of body fluids, as well as electrolytes. The body of a young calf contains 75% water and a loss of 10% puts its life in danger, while a loss of 15% results in death. Sunken eyes are one symptom of dehydration, which, if advanced, will cause the upper eyelashes to be directed towards the inside of the eye socket, obscuring the calf's vision. The level of dehydration can be checked by pinching a bit of skin near the ribs and twisting it 90 degrees. The slower the fold of skin springs back to its original position after release, the higher the level of dehydration and the quicker the need for treatment. See Table 11.1 for further details.

- **The calf has lost hair around its muzzle and/or rectum and along its hind legs.** Offering hot milk to the calf or letting the manure stick to the skin for a long time are often stated as the main reasons for loss of hair. If these two causes can be excluded, poorly emulsified fat in milk replacer is a likely suspect. Fat globules attach themselves to the skin and prevent the air reaching the hairs. Similarly, hair is lost around the rectum when it is in contact with the undigested fat in the manure. Low digestibility of fat in milk replacers containing high levels of non-clotting plant protein may have the same consequences. If the flesh is raw, wash it with a clean cloth, wrung out with soda water. The whole area should be treated with a weak solution of iodine.
- **The calf bloats after drinking milk.** Under certain situations, the oesophageal groove does not close completely, thus leaking milk into the rumen. This can occur through rough handling, feeding milk that is too cold or too hot, overfeeding or force feeding when the abomasum is not sufficiently empty. It can also occur when the calf is sick or when fed poor-quality milk replacer. Feeding milk through rubber teats, or at regular intervals, at body temperature, and in small quantities may help re-establish the proper function of the oesophageal groove. Letting the calf suck your finger for a moment before offering the milk bucket will also help.
- **A weaned calf bloats on *ad lib* grain feeding.** The sudden accumulation of gas in the rumen that cannot be expelled can even occur in calves that are well adjusted to high-grain diets. Within 1–2 hr after feeding, the left flank rises very quickly, the calf nervously lies down and tries to defecate. If only one or two calves bloat, then it is unlikely to be due to the feed or feeding practices. Some calves are just prone to bloat and will get over it without any treatment. Regrouping calves may allow previously submissive calves better access to the grain, which can upset rumen gas expulsion if this happens too quickly.
- **The calf has a foul-smelling greenish liquid dripping from its mouth and it loses its cud.** This is sometimes called 'medicine disease' and is caused by prolonged use of antibiotics, which upset the balance of rumen microbes. The best option is repeated introduction of a cud from a healthy animal, preferably on the same diet, into the sick calf. A similar effect, called 'microbe swapping', can also occur during hand feeding of calf starter in newborn calves. Some rearers consider that the dripping is caused by calves twisting their heads to the side while drinking from rubber teats. This can cause the oesophageal groove to malfunction, thus allowing milk to enter the rumen and upset the establishment of normal populations of rumen microbes. Another possible cause is damage to a large portion of the rumen wall by prolonged scouring or the presence of small pieces of wire.
- **The calf develops a pot belly.** This indicates a long-term nutrient imbalance, in that there is too much fibre and too little energy in the diet. High-fibre diets require high water intakes, which together with the slowly digested feed increase rumen volume. Energy intake is further reduced through a limited gut capacity and this leads to poor growth. Sometimes pot bellies develop in calves suffering from internal parasites, those with a damaged gut from chronic scouring or those with a long history of pneumonia. The obvious treatment is to feed more energy and less roughage. By

feeding *ad lib* concentrates and a low-quality roughage, calves will only eat about 10–15% of their diet as fibre and the rest as concentrates.

- **The mouth cavity and skin under the calf's eyelids are pale.** Calves fed milk exclusively will show signs of anaemia, due to the low iron levels in the diet. Once eating solid food, this problem will disappear as concentrates contain sufficient iron for calf requirements. If growing calves for white (milk-only) veal, intake and performance can suffer through anaemia. This can be corrected, without endangering the marketability of calves, with intramuscular injections of iron.
- **The calf has a dry, hot muzzle.** This calf would have a high body temperature and most likely be suffering from a respiratory disorder. Electrolytes and antibiotics would probably help.
- **The calf has a nasal discharge.** A transparent, watery discharge indicates the calf is, or was, exposed to significant environmental, housing or nutritional stress. The cause of the discharge is usually a viral infection, like a human cold. Remove the source of stress and if the body temperature is elevated offer three adult size aspirins per day. If the colour of the discharge changes to brown or greenish and is thicker, then the body is already fighting a secondary bacterial infection.
- **The calf's temperature dropped after treatment but rose again a few days later.** This could be due to several possibilities:
  - The correct treatment was applied but for too short a period.
  - The treatment was applied once each day whereas twice-daily treatment would have provided better uniformity of antibiotic release.
  - The level of drug applied was insufficient.
  - A combination of the chosen antibiotic along with an anti-inflammatory drug was administered and the temperature drop was solely due to the anti-inflammatory drug, which may have masked the improper selection of the antibiotic.

  It is important that the veterinarian should identify the disease organism responsible for the stress to ensure the most appropriate treatment can be given.
- **The calf has saliva running from its mouth.** This could indicate many disorders, but is mainly connected to severe pneumonia. The front legs are spread, the neck is stretched, the head points to the ground and breathing is laboured. Saliva will be running either in a clear steady stream or as a slow-moving liquid. In many cases, even dramatic measures cannot save such a calf. Move it to a well-ventilated isolation pen, provide good bedding and fresh feed and water. Veterinary attention is essential.
- **The calf has an umbilical hernia.** Hernia or rupture is a protrusion of one or two loops of intestine or other tissue from the abdominal cavity through the navel opening. If such an opening is no more than 2.5–4 cm, it usually closes sufficiently when the calf grows older. Larger openings require surgical correction. Taping the opening for a period of 4 weeks may be necessary if the hernia is two fingers wide at 2–3 months of age. Application of rubber rings (used for tail docking in lambs) to the skin pouch only are effective in heifer calves. Use of more than one rubber ring prevents them from sliding down. The rings stop the blood supply to the navel and in 2 or 3 weeks the navel cord will fall off and the connective tissue will close the opening.

- **The calf has warts.** Warts are a specific skin overgrowth caused by a viral infection. In calves they sometimes appear on the head, the ears and around the mouth and eyes. They are contagious to other animals and some can even be transmitted to humans.
- **The calf has manure accumulation around its hooves.** Dry clusters of manure can have very unpleasant consequences. They can cover an infection, be filled with fly maggots or lead to abnormal wear of the hoof. The feet should be checked at regular intervals using a blunt edge of a putty knife to remove the manure between and around the hooves. If the skin under the removed manure is red, mouldy or smells, wash it with diluted iodine.
- **The calf's mouth is cold.** You are losing this calf. The body defences are breaking down and infection is taking over. The body temperature is well below normal, usually below 35°C, and the chances of recovery are very slim. In an attempt to raise its temperature, try thick, dry bedding, or plastic bags filled with warm water or heat lamps while lukewarm milk and/or water could be offered. Do not raise your hopes too high.

12.5 Understand how calves react to people

How much do we really know about the basic sight and hearing senses of calves and heifers? An article by two US calf-rearing specialists (Leadley and Sojda 2001) tells us much of which we may take for granted, but, on the other hand, may not even be aware of. Firstly, cattle have wide angle vision: they can see 300 out of 360 degrees around them. They use this field of vision to define their 'personal space', which we call their 'flight zone'. Secondly, cattle are quite sensitive to high frequency noises, and compared with people (who can hear noises from 1000–3000 hertz), they can hear noises up to 8000 hertz. The authors have listed some general rules to help with cattle handling:

- When a person moves into their flight zone, cattle will normally try to move away.
- The size of their flight zone will decrease slowly if they are handled frequently and gently.
- Previous experiences will affect how animals react to future handling, with memories persisting for many months. Obviously memories involving fear are significant in increasing flight zones.
- Calves can readily tell the difference between two situations and make choices to avoid the more stressful one.
- Cattle are sensitive to changes in colour and texture.
- Moving objects and people seen through sides of a chute can frighten animals.
- Novelty is a strong stressor, while repeated exposure will reduce the novelty effect.
- Cattle are herd animals and do not like to be separated from their herd mates.
- Groups of cattle that have body contact remain calmer.
- Unexpected loud or novel noises can be highly stressful.
- Cattle readily adapt to reasonable levels of continuous sound such as background noises or music.

- Cattle exposed to a variety of sounds, such as radios with talk and music, may have a reduced reaction to sudden noises.
- Cattle readily adapt to handling, even if the events may be initially stressful, such as walking up a race, into a head bale or being transported.
- Cattle can be trained to voluntarily accept restraint with relatively low levels of stress.
- A small amount of inconsistency in care and handling can reduce calves' stress response to new sights and sounds.
- Consistent poor handling can create chronic stress.

Calves require handling techniques different to those used with adult cattle; hence, an experienced handler of adult cattle will not necessarily have the skills to handle calves. Calves spend 90% of their time lying down and sleeping and do not have herding and following behaviour of adult stock. They do not have a flight zone and they are often not fearful of humans. In fact, they are more likely to move towards people because they associate them with food. They move independently of each other and their movements can be unpredictable. In addition, calves are inquisitive and can be easily startled by noise.

Calves must never be:

- thrown, dropped or dragged
- struck, hit, punched or kicked
- shocked with an electric prodder
- moved by dogs.

Because calves respond to a quiet and reassuring voice, calf handlers should move in a slow and predicable manner. Calves should be handled patiently and quietly to reduce stress and risk of injury. A high degree of kindness and empathy is needed when working with calves to reduce stress and risk of injury.

These basic rules can partly explain why empathetic calf rearers do a good job, whereas insensitive rearers do a poor job. Just spending time with young calves, particularly newborn ones, develops that essential bond, while quiet consistency in all management procedure, even to the point of clothes worn in the calf shed, ensures the calf nursery is as peaceful as any infant's bedroom.

12.6 Communicate with your calf rearer too!

Farm managers and other employers of farm staff should be aware that praise is one of the best motivators for employed labour. When your calf rearer does a good job, be certain to say frequently out loud and face to face, 'Thanks for doing a good job!'

Proficient calf rearing requires use of the fives senses (sight, smell, hearing, touch and even taste), and this takes time to develop. Calf rearers have to be extra alert and ready to act quickly when a calf is ill. Timely diagnosis and treatment are measured in minutes rather than days. This kind of care calls for lots of flexibility and commitment on the part of the calf rearer. Good calf rearers have a bond with their calves that is tied to this commitment.

Giving that 'little extra' over and over again, week after week, is costly for the rearer. It means being continually alert when working with the calves, so sick calves are quickly identified and treated. It may also mean returning to the calf shed at night to administer antibiotics or electrolyte fluids. Like all aspects of dairy farming, flexibility is a key attribute in calf rearing, where all staff must prepare for the unexpected. We all like to work to routines, particularly employed labour, but all too frequently emergencies can occur and daily work schedules must be quickly modified.

Owners and managers must allocate sufficient labour resources at the right time and place; for example, when assisting in calving, time should be allowed for dipping the navel with iodine and dosing newborn calves with colostrum. Farmers should also provide opportunities to spread out stressful events rather than stack them one on top of another. For example, vaccinations, dehorning, tail docking, ear tagging and weaning are all stressful to the calf, and even to the rearer if it requires continually handling and restraining calves.

When selecting or constructing rearing facilities, rearers should also be kept in mind. Ensure that staff can easily see all the calves during a single patrol down the calf shed. Provide enough hot water for cleaning buckets, teats and other feeding equipment, and cold water outlets from which calves can drink. Importantly, provide good quality and palatable feeds, such as concentrates, roughage and milk replacer. If calves quickly develop a taste for solid feeds, they require less labour input and are less likely to suffer ill health.

Skilled, motivated and empathetic staff are a major contributor to a successful calf-rearing operation. However, because no one is born with such skills, they must be learnt through experience. Therefore it should be assumed that, until a farm worker can successfully demonstrate a particular skill, he or she does not have it. In teaching any new farm practice to workers, explain why it is done this way so they can better understand the skill, and then ask them to repeat it, showing that they can successfully do it in the way they were shown. Remember that learning a new skill depends on being shown how to do it correctly.

12.6.1 Developing standard operating procedures

A standard operating procedure (SOP) is a formal term used to describe a set of instructions for any activity or set of tasks undertaken in the workplace. It should be described in sufficient detail that any non-skilled person could attempt to undertake it. Such SOPs can be developed in such as a way as to undertake the most suitable procedures to achieve the most desirable, or best possible, outcome, in which case they become best management practice (BMP). A series of BMPs are presented in Chapter 18 of this manual to describe the entire range of activities involved in calf and heifer rearing.

It is important to develop SOPs for major farm tasks because on large farms they are likely to be carried out by more than one farm worker. Calves, as well as farm managers, perform better when all farm practices are undertaken in a routine manner. The benefits of SOPs include:

- They lead to consistency even when undertaken by different people.
- Calves perform better if routines do not change.
- People thrive on consistency as they know exactly how to do any job and what the outcome should be.

- Training new staff is easier because there are a series of carefully documented steps to achieve that outcome.
- Because they are written down, any member of the farm staff can refresh their memory on how to perform the tasks.
- They can be referred to by managers to follow through any farm activity that did not give the desired outcome.
- They can even be used as legal documents in the case of formal disputes.

Generic SOPs can be developed for any set of tasks, but the best ones are those developed by the staff (workers and management) actually doing it, because they are most familiar with the farm operations and the infrastructure, equipment and methods used on that farm.

Fisher (2009) describes a simple approach to developing an SOP, as follows, **using the example of maximising passive transfer of immunity through colostrum feeding**:

- Prioritise the areas that would benefit from an SOP, namely those that would most benefit from a series of clear written protocols [**calving down cows and heifers**].
- Select the most appropriate farm staff to oversee the development of the SOP, obviously those with overall responsibility for that task [**colostrum feeding and milk feeding**].
- Ensure that anyone likely to undertake this task is involved in its development [**other farm staff involved in calf rearing and animal health**].
- Make a list of processes, within the selected area [**such as colostrum collection and quality assessment, storing colostrum**].
- The extent of what the SOP covers and what is does not [**does it cover cleaning and sanitising feeding equipment, does it cover cleaning udders?**].
- Give the SOP a specific name [**maximising the passive transfer of immunity**].
- Detail its scope [**as above**].
- Prominently list hazards that exist and precautions that should be taken [**None**].
- Detail any safety equipment or protective clothing required [**None**].
- List all equipment and supplies needed [**buckets, colostrum quality measuring equipment, stomach tubes or nipple bottles**].
- Detail in sequence, the steps needed to be taken to achieve the desired outcome [**these can be summarised from Chapter 5 of this manual**].

Once completed, the SOP can be briefly summarised and placed on the wall of the calving down pens. The more detailed SOP should be located in the farm office and also the staff quarters. This should be reviewed say every 12 months and updated to remain pertinent to the task.

12.7 Contract calf rearing

Dairy farming is becoming a specialist profession requiring many skills. Rather than keep up to date with all these skills, increasing numbers of farmers now outsource particular enterprises on their farms. Contractors now offer their services in forage conservation, milking, heifer rearing and, more recently, calf rearing. Skilled rearers collect newborn heifer calves from the farm, milk rear them and return the weaned calf at about 12 weeks of age, weighing 100 kg. The rearer is often provided with transition milk with which to commence milk feeding, but then the diet is changed to milk replacer. Early weaning, at say 5 weeks, would reduce total feed costs.

Recipient farms have to be selected carefully to minimise the introduction of diseases, thus reducing the rearer's concern about spreading diseases among the contracted calves. They should draw up formal contracts, or at least agree on the costs of disease treatment and mortalities, and target live weights. Rearers generally bulk purchase milk replacer and calf pellets, so they often develop good alliances with manufacturers.

In some instances, calf rearers contract to rear the calves on the home farm, thus providing only the labour and expertise, while using all the farm facilities, including transition and vat milk.

There may also be a role for the specialist farm midwife, whose job it is to routinely check the springing cows and provide assistance with calving. They can assist with natural suckling of newborn calves or remove them at birth to sheltered pens, and artificially feed them their first colostrum. Such a position may need only work for several months each year and that person could contract to do it on several farms in close proximity. The costs involved in employing such a farm midwife would be offset by the reduced time spent with the calving cows and their progeny (particularly late at night). There would also be the added benefit of reduced health problems and mortalities arising from a guaranteed higher level of passive immunity among the replacement heifers. The details of contract rearing of weaned heifers are discussed in Chapter 13.

12.8 Codes of animal welfare for calves

Throughout the world, public perceptions of farm animal welfare issues have the potential to affect the sustainability of livestock industries markedly, with national and international pressures likely to have increasing roles in determining how animals are managed. Because farm animal welfare is largely part of good animal and farm management, paying close attention to their day-to-day management should also ensure acceptable welfare.

Calves born to dairy cows are routinely submitted to more insults to normal development than any other farm animal. They are taken from their mother generally within their first day of life, often not even allowed to suckle colostrum immediately after birth, and are frequently deprived of their most natural feed: whole milk. They may be fed one of the varieties of cheaper liquid substitutes for milk, but, because these are still more expensive than solid foods, many are weaned off milk as quickly as possible.

Of all the classes of dairy stock, the welfare of milk-fed calves has received the closest scrutiny. The following summarises the current Australian codes for their health welfare and wellbeing (Moran 2002).

12.8.1 Housing

- Housing should be hygienic, with adequate ventilation, climate control and lighting. Flooring should be well drained with adequate dry lying space for each cow. Flooring and internal surfaces should not cause injury and should allow easy cleaning.

- Calves should be given access to sufficient bedding or appropriate flooring that ensures comfort and a clean, dry place to lie.
- Feeding systems should be designed to permit easy access and reduce bullying.
- The floor area must be sufficient to enable each calf to freely turn around, stretch and lie down comfortably. A floor area of at least 1.5 m^2 should be provided for each calf individually housed in pens or cribs. Pen heights should be a minimum of 1 m with provision of additional height to allow for ventilation space.
- Calves, being social animals, seek the company of other calves, so should be able to see other calves during early rearing.
- Where large numbers of calves are reared, they should be grouped by age and size to reduce competition for food and to allow closer observation and management.

12.8.2 Feeding

- Calves require at least 2 L of fresh or preserved colostrum within the first 12 hr following birth. Calves should continue to receive colostrum for the 3 days after birth.
- Thereafter, they should be fed at least daily on liquid milk, or milk replacer, in sufficient quantities to provide essential requirements for maintenance and growth.
- Hay, calf concentrate or high-quality forage should be available to calves no later than 3 weeks of age to help in the development of their digestive tract and to ease the stress of weaning.
- Calves should be weaned off milk or milk replacer onto rations providing all essential requirements only when their ruminant digestive systems have developed sufficiently to enable them to maintain growth and wellbeing.

12.8.3 Management practices

- Restraint should be the minimum necessary to perform management procedures efficiently.
- Procedures and practices that cause pain should not be carried out if painless and practical methods can be adopted to achieve the same result.
- Special care should be given to the practices of castration, tail docking (if undertaken) and dehorning.
- Any injury, illness and distress should be promptly treated.
- Appropriate preventative measures should be implemented for diseases likely to occur in the herd. A suitable vaccination, and internal and external parasite control plan should be devised and followed for each farm.
- Internal medication, such as vaccines and drenches, and external medications, such as dips and pour-on formulations, should be stored and given in strict accordance with the manufacturer's instructions and recommended methods of administration. Overdosing may harm cattle, while under-dosing may result in failure of the medication. Expiry dates and withholding periods should be strictly observed.
- The preferred method of humane destruction (euthanasia) of cattle should be by overdose of anaesthetic or using a gunshot.
- Killing day-old calves may also be achieved by a heavy blow to the crown of the head to stun the calf prior to bleeding out.

12.8.4 Transportation of calves

- Transportation of stock should ensure that they reach their destination as speedily as possible, within the confines of the road laws, and in a condition not significantly less than when assembled for loading. The possibility of either injury or illness during transport should be reduced to a minimum.
- Handling of calves should be carried out in a manner that will avoid injury or unnecessary suffering. Calves are not to be kicked, beaten, pulled, thrown or prodded with any sharp instrument. The use of electric goading devises or dogs when handling, driving, drafting, weighing, loading or unloading calves is not an acceptable practice.
- Facilities should be constructed to permit safe loading and unloading of calves.
- Places where calves are held should have facilities and/or contingency plans to feed calves in the event of delayed removal or slaughter.
- Calves should be fed at least every 24 hr and have access to drinking water.
- The driver of the vehicle is responsible for the care and welfare of all animals during transportation.
- Animals that either become ill or weak, or are injured during transport should receive appropriate attention and treatment; if necessary, they should be slaughtered humanely.
- Vehicles used for transportation should be thoroughly cleaned prior to loading and at the end of the journey.
- Transport operators should check calves en route at least every 3 hr.
- Calves should be loaded at a density so as to allow all calves to lie down while being transported.

13

Post-weaning management of dairy heifers

This chapter discusses the benefits of well-grown dairy heifers and the growth targets and feeding programs to produce them.

The main points in this chapter

- All too often dairy farmers do a good job with rearing heifer calves up to weaning then virtually neglect them thereafter.
- The rearing of heifers can be divided into two stages: from weaning to first service and from first service to calving.
- Undersized heifers have more calving problems, produce less milk, have greater difficulty getting back into calf and compete poorly with older cows for feed.
- Because they are still growing, heifers will use some of their feed for growth rather than for producing milk and are more likely to be culled for poor milk yield and/or infertility.
- There are many published targets for pre- and post-calving live weights (hence growth rates), body condition, wither heights and first lactation milk yields, but these have all been developed for temperate dairy production systems. There is little information on tropical dairy systems.
- There is little doubt that heifers will not achieve the same growth performance in the tropics as they would in temperate dairying areas, although they may achieve similar mature live weights.
- Realistic in-calf target weights for first calf dairy heifers on tropical dairy farms should be of the order of 350 kg (for Zebu and crossbreds) to 450 kg (for grade Friesians). However, from anecdotal evidence, most tropical dairy farmers are not achieving such targets.

- To accurately quantify weight for age, it is important to be able to assess age from the eruption of incisor teeth.
- Providing weaned heifers with a diet that is both short on daily DM intake, as well as being low in energy content, is a major reason for sub-optimal heifer growth rates. This leads to light-weight heifers and delayed calving, hence reduced potential to produce milk and to get back in calf.
- Heifer farms are a new initiative in many Asian countries in which week-old calves are collected from individual farms and group reared in one location prior to their return to that farm just prior to calving down.

This chapter discusses the post-weaning management of calves reared as replacement heifers for dairy farmers, as summarised in Figure 13.1, as well as a small section on producing beef from the dairy herd.

The rearing of heifers can be divided into two stages: from weaning to first service and from first service to calving. The aim of heifer rearing should be to achieve the maximum growth and development, arriving at the earliest sexual maturity consistent with the least cost. This ensures that maintenance costs are minimised, that there is the earliest possible return on the investment in the original animal and that the heifer can produce well during her first lactation.

13.1 On-farm rearing of replacement dairy heifers

All too often, dairy farmers do a good job of rearing heifer calves up to weaning, but then virtually neglect them thereafter.

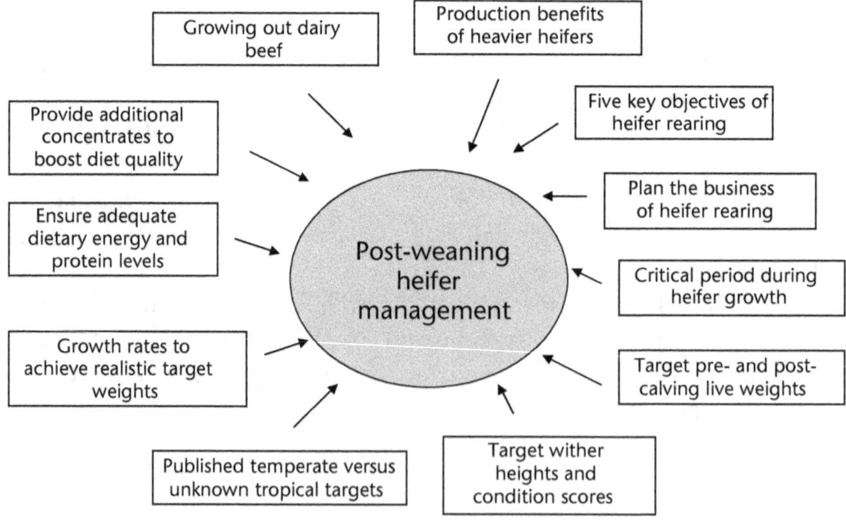

Figure 13.1 The key factors to consider in post-weaning management of dairy heifers

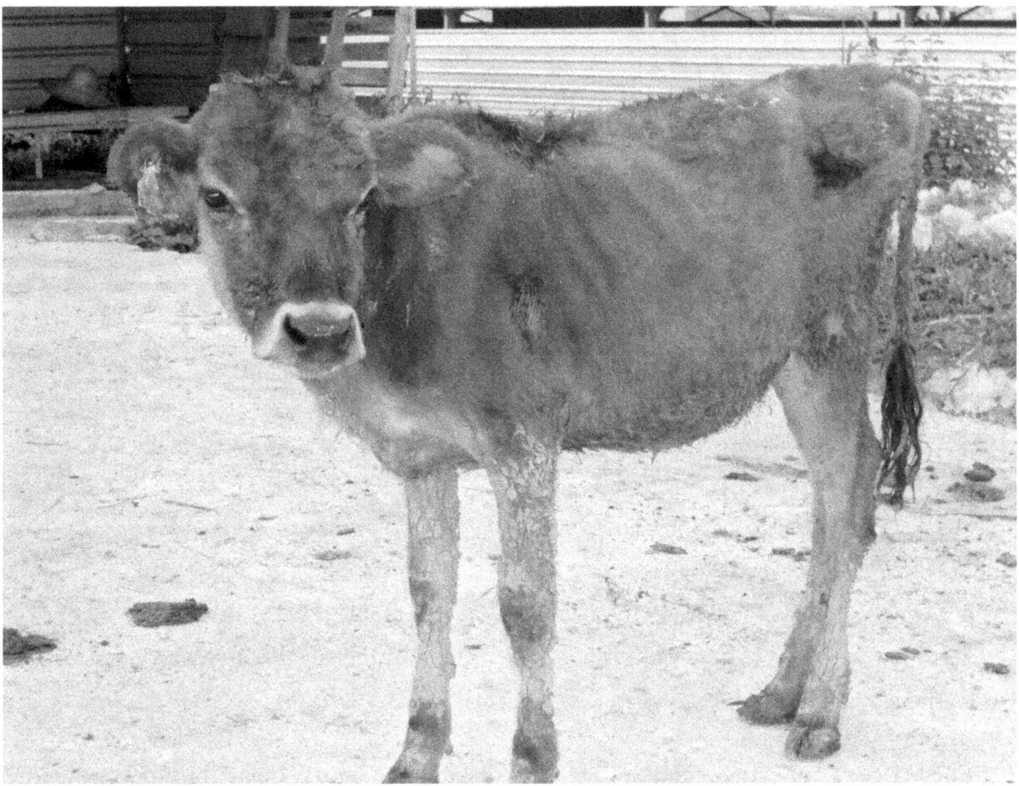

A very poorly reared Jersey heifer in Malaysia.

Weaned growing heifers require less attention than milk-fed calves and milking cows. From weaning until breeding and sometimes even after then, daily contact is not necessary. Because their nutrient requirements are relatively low compared with milking cows, many heifers are located away from the prime grazing areas on many dairy farms, sometimes on agistment on other farms. Unfortunately, the saying 'out of sight, out of mind' applies too frequently to replacement heifers. This relative neglect is understandable in view of the long time it takes before any inadequacies in post-weaning practices are reflected in poor milking cow performance.

Dairy heifers need to be well fed between weaning and first calving. Growth rates should be maintained, otherwise heifers will not reach their target live weights for mating and first calving. Undersized heifers:

- have more calving difficulties
- produce less milk
- have greater difficulty getting back into calf during their first lactation
- when lactating, they compete poorly with older cows for feed
- because they are still growing, will use some of their feed for growth rather than for producing milk
- are more likely to be culled for poor milk yield and/or infertility.

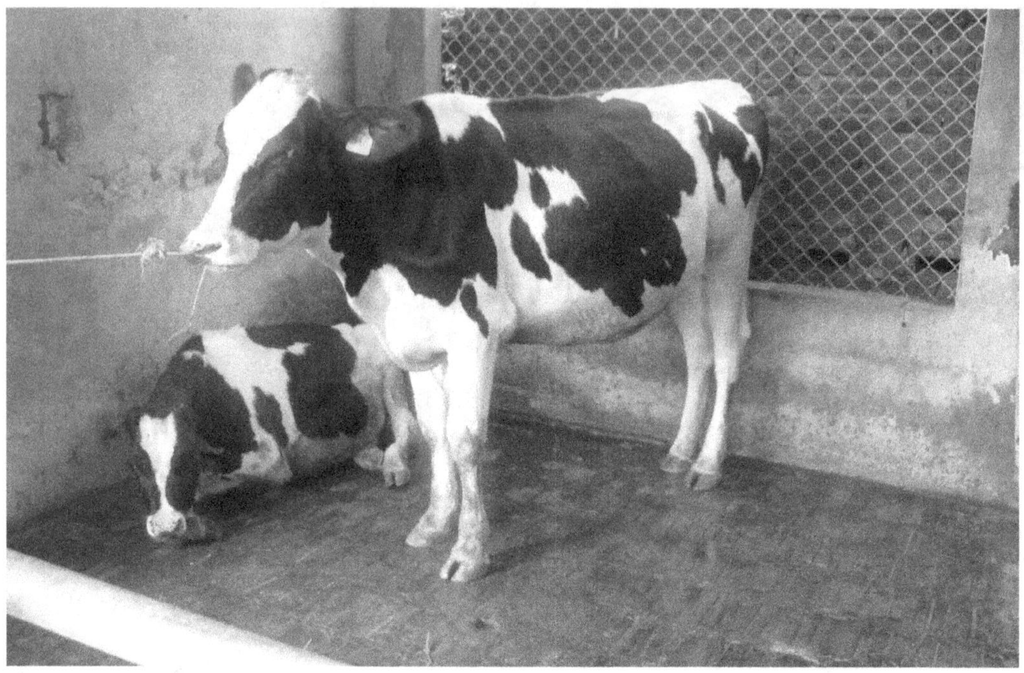

A well-reared pregnant Friesian heifer in Vietnam.

The onset of puberty is related to weight rather than age. A delay in puberty means a later conception, which can disrupt future calving patterns and increase rearing costs. All heifers should reach a minimum weight before joining, because lighter heifers have lower conception rates. Target live weights at mating and first calving are discussed below.

To recommend detailed post-weaning management procedures for dairy heifers is not practicable. The system adopted should reflect local and climatic conditions and personal preferences. The extremes of weather and availability of home-grown and purchased feeds are probably the most important variables. Although replacement heifers are essentially non-productive animals, some expenditure is necessary. They represent capital and investment in the dairy herd's future. Heifer rearing should achieve the maximum return on this investment with a minimum of outlay. It should not be regarded as a haphazard undertaking, which hopefully will produce a pregnant heifer, but rather as a business enterprise with clearly defined goals such as:

- the number of animals to be reared
- the desired age at first calving
- the target live weight at calving
- the feeding program
- ways to monitor their performance and total rearing costs
- any specific housing and health requirements.

When rearing dairy replacement heifers, producers should have five major objectives:

1. **The maintenance or expansion of herd size.** Heifer-rearing systems should provide sufficient animals to replace cows culled from the milking herd and allow for increases in herd numbers, if required.
2. **Calving by 24–30 months of age.** Entry into first lactation by 24 months of age minimises the total non-productive days and maximises lifetime productivity.
3. **Sufficient growth for minimal dystocia (that is, calving difficulties) at first calving.** Heifers need to be large enough to calve without difficulty.
4. **Maintenance of health.** The prevention of clinical and sub-clinical disease plays a large role in the ability of replacement heifers to meet live weight and age targets at first calving. Longevity and lifetime productivity is also affected.
5. **Genetic progress.** Replacement heifers generally have higher genetic merit than the current milking herd. This can be expressed as increased productivity (both milk volume and solids), improved efficiency of production and/or enhanced resistance to disease.

When considering these objectives, producers should decide whether to rear their own replacements on farm, to have them contract reared off farm or to purchase in-calf heifers. The last two alternatives will save land for milking cattle, which is important where land is the major constraint to production, but are likely to cost more than on-farm rearing. Contract heifer rearing is discussed in a later section in this chapter. If the farm has a history of poor calf- and heifer-rearing practices leading to high mortalities and poor heifer growth rates, off-farm rearing or purchasing of in-calf heifers should be a serious consideration. However, it must be remembered that when purchasing in-calf heifers, there is no guarantee that their genetic merit is superior to that of older cows in the herd, while their health status is largely unknown.

Heifer rearing is not cheap and the costs to produce a lactating first-calf heifer can account for 15–20% of the total milk production costs. It is not good economics to cut back on heifer rearing costs because lifetime profits will be reduced (Moran and McLean 2001).

The number of first calving heifers each year will depend on the replacement rate within the milking herd. This is the sum of the wastage rate caused by infertility, mastitis, low milk yield, old age, accidents, and so on, together with the particular culling policy for that herd, whether this is to improve milk yield, feed efficiency or reduce calving interval. The number of heifer replacements to be reared also depends on mortality rates during rearing, the conception rates at first mating and the proportion of heifers reaching their target weight for age.

With 20–30 heifers per 100 cows introduced into the milking herd annually, at least 80% of the milking cows should be mated (either using a bull or artificially insemination) to obtain that number of replacements each year. When determining the total number of calves to rear, consideration could be given to rearing additional heifers for sale to other dairy farmers and/or bull calves for dairy beef.

On well-managed farms in temperate dairy regions, achieving a consistent calving program requires heifers to:

- reach puberty at about 12 months of age
- become pregnant at 14 or 15 months of age

- calve at 24 months of age
- return to oestrus and be mated within 70–80 days of calving.

Earlier first-calving ages are easier to achieve with the smaller, more rapidly maturing dairy breeds such as Jerseys, Ayrshires or Zebus.

13.2 Benefits of heavier heifers

Provided heifers are at least 18 months old, the younger the heifers calve, the higher their first lactation and mature milk yields, the more calves they produce and the longer their productive life in the herd. An additional benefit is a more rapid generation interval, and hence a faster rate of genetic progress in the milking herd. Lifetime productivity reaches a peak in heifers calving at 25–27 months of age. For example, heifers calving at 24–27 months of age can produce 21 000 L of milk over a 7 yr productive life on established temperate dairy farms, compared with 18 750 L if calving at 30–33 months and only 17 000 L if calving at 36–42 months of age.

Several studies have documented the long-term benefits from heavier calving weights in Friesian heifers. For every additional kilogram at first calving, heavier heifers can produce 7 L of extra milk in each of their first three lactations. Therefore, if heifers calved at 500 kg compared with 450 kg, they would produce an extra 350 L of milk/lactation, or 1050 L extra milk over their first three lactations.

As part of an Australia-wide survey of dairy herd fertility, involving over 33 000 cows, live weights of 2000 Friesian heifers from 69 seasonal calving herds in Victoria and Tasmania were recorded just prior to their first calving (John Morton *pers. comm.*). Heavier heifers calved earlier and conceived more readily during their first lactation (see Table 13.1), indicating a decreased need to induce (or even cull) second calving cows.

Heavier calving live weights also reduce the incidence of calving difficulties and wastage rates, which both adversely affect lifetime performance and herd profits. The percentage of replacement that either die or are culled before their second calving has been recorded at 30–35% in two Victorian studies, which is considerably higher than the target 20% possible with better grown heifers. Similar data for the tropics is not available.

Table 13.1. The effect of live weight at first calving (LWFC) on percentage of Friesian heifers calving in their first 3 weeks of the calving period, subsequent 3 week submission rate, 6 week in-calf rate and the proportion of heifers conceived between 7 and 21 weeks during their first lactation

LWFC (kg)	Calved in first 3 weeks (%)	3 week submission rate* (%)	6 week in-calf rate* (%)	Heifers conceived from 7 to 21 weeks of mating* (%)
<400	36	58	49	30
400–440	49	74	60	27
440–480	55	77	68	21
480–510	65	82	68	19
510–540	53	85	75	13
>540	68	88	77	10

* Data are expressed as percentage of first-calf heifers mated.

There is a critical period for the developing udder during which time excessive growth rates can be detrimental to lifetime productivity due to increasing the deposition of fatty tissue in the udder. Exactly when this critical period occurs and exactly what constitute excessive growth rates have yet to be clearly defined, although there are some general guidelines. Live weight gains should not exceed 0.8 kg/day between 6 and 12 months. Hence heifers should not be fully fed during their second 6 month period. This is unlikely to be a problem in the tropics where post-weaning feeding regimes would rarely lead to such high growth rates.

13.3 Targets for growing heifers

13.3.1 Target pre-calving live weights

Table 13.2 summarises the target live weights for Friesian and Jersey heifers at various ages on well-managed Australian dairy farms (Moran and McLean 2001).

Unless greater attention is given to heifer feeding, most tropically reared heifers are too small, hence too sexually immature, to breed at similar ages as their temperate counterparts. It is better to use target live weights rather than ages to plan heifer mating programs. Adequate breeding weights in the tropics would then be 200–220 kg for small breeds and 290–310 kg for larger breeds.

There is little doubt that heifers will not achieve the same growth performance in the tropics as they would in temperate dairying areas, although they may achieve similar mature live weights. The major constraints to heifer performance in the tropics are heat stress, in both indoor- and outdoor-raised heifers, and forage quality. Chapter 6 provides details on why the tropics is not an ideal place to rear dairy calves and heifers.

13.3.2 Target in-calf live weights

Heifer milk production depends on their live weight at first calving and how well they are fed and managed as milkers in their first lactation. Their optimum live weight at first calving (LWFC) depends on the milk yield farmers wish them to achieve at maturity in the herd. Table 13.3 presents data on target live weights (in-calf) for 2-yr-old Friesian

Table 13.2. Target live weight ranges (kg) at various ages for well-managed Friesian and Jersey heifers on Australian dairy farms

Age (months)	Friesian	Jersey
3 (fully weaned)	90–110	65–85
6	150–175	110–130
9	210–235	155–180
12 (yearling)	270–300	200–230
15 (mating)	330–360	245–275
18	390–420	290–320
21	455–485	335–365
24 (pre-calving)	520–550	380–410

Table 13.3. Target live weights for 2-yr-old Friesian heifers to enable them to produce a specified milk yield as mature cows

Full lactation milk yield as mature cows (litres)	Target live weight for 2-yr-old heifers (kg)
3000	430
6000	540
9000	590

heifers required to produce a subsequent full lactation yield as mature cows (Moran and McLean 2001). On most well-managed temperate farms, 6000 L of milk/lactation would be a realistic target, meaning that Friesian heifers should be grown out to 525–550 kg as 2-yr-olds just prior to calving. The extremes in this table represent typical values for average milk yields on many SHD farms in the tropics (3000 L) and on intensive temperate dairy feedlot farms (9000 L).

The target weights in Table 13.3 are for purebred Friesians, so would be smaller for crossbred Friesians or Zebu (tropically adapted) dairy heifers. Furthermore, they are for 24-month-old heifers, whereas corresponding LWFC would be older, say 30 months of age. Therefore I believe that corresponding target weights for in-calf heifers on tropical Asian dairy farms should be of the order of 350 kg (for Zebu and crossbreds) to 450 kg (for grade Friesians). From anecdotal evidence, most Asian dairy farmers would not be achieving such targets.

According to US standards (BAMN 2007), the mature weight of a dairy cow is its live weight during its third lactation, generally measured during mid lactation and/or when cows have a good body condition score (six out of eight points). BAMN (2007) then presents target weights expressed as a percentage of mature weights, these being:

- 45–50% at puberty
- 55% at mating, generally during their third oestrus after puberty
- 82% after first calving
- 92% after second calving
- 100% after third calving.

Table 13.4 presents the mating and calving targets for dairy stock of different mature live weights. It should be pointed out that these weights are after calving, whereas those in Table 12.2 includes the weight of the foetus and associated amniotic fluid and associated tissues, say 50–70 kg.

Table 13.4. Target live weights (kg) for dairy heifers and cows of different mature live weights

Stage of development	Percentage of mature weight	Target live weight (kg)				
		400	450	500	550	600
First mating	55	220	247	275	302	330
First calving	82	328	369	410	451	492
Second calving	92	368	414	460	506	552
Third calving	100	400	450	500	550	600

Tables 13.2, 13.3 and 13.4 are recommendations for temperate dairy systems. Until such data has been generated for tropical dairy farms, these are really the only guidelines we have on which to base post-weaning heifer feeding management in Asia. They should be achievable on well-managed tropical dairy systems, but would be overoptimistic estimates on most small holder farms.

13.3.3 Target growth rates

When calculating the desired growth rate to achieve these targets, consideration must be given to the current live weight, the target weight and the time period over which the weight gain must be achieved. Four examples are presented in Table 13.5 to achieve target mating weights, for heifers with different mature live weights and current weights for age.

13.3.4 Target chest girths and wither heights

Chest girths, which measure the linear distance around the heifer's chest, can be used to estimate live weights: relationships are presented in Moran and McLean (2001).

Wither height (or height at the shoulder) is a good measure of bone growth in heifers, and therefore frame size. Frame size can influence the area between the pelvic bones, and hence ease of calving. It can also influence the abdominal capacity, and hence the appetite, of milking cows. Wither height may even be a better measure of heifer development than live weight, because this is influenced by pregnancy and body condition. It is measured as the highest point on the heifer's shoulder, immediately above the front legs and can be routinely recorded by locating a marked stick in the crush or raceway to be read as the heifers move past. Animals should be standing quietly, but not with their head in a head bale, because this alters their natural stance.

Target wither heights in Friesians are 123–125 cm at 15 months and 133–135 cm at 24 months. Corresponding wither heights in Jerseys (and Zebus) would be 110–112 cm at 15 months and 120–122 cm at 24 months.

A heifer's appearance, condition and coat are useful indicators of her health and performance. A heifer with a solid slab-sided appearance and a sleek coat is obviously doing well. Conversely, an animal with a 'gutty' appearance and with more prominent backbone and hips reveals inadequate energy in the diet, such as young animals fed mainly on pasture or low-quality roughages with insufficient concentrate supplements. The coat of poorly grown heifers may appear more coarse and hairy. These symptoms can also indicate a high internal parasite burden.

Table 13.5. Target growth rates (kg/day) for dairy heifers of different mature live weights and various current weights and ages at first calving to achieve target mating weights

Mature live weight (kg)	Age at first calving (months)	Current age (months)	Current weight (kg)	Target mating weight (kg)	Target mating age (months)	Target growth rate (kg/day)
450	24	4	100	247	15	0.44
500	24	4	120	275	15	0.47
500	26	4	120	275	17	0.40
500	28	4	120	275	19	0.34

Body condition should also be taken into account, as well as live weight and frame size, because over-conditioned heifers are harder to inseminate and are at greater risk of calving difficulties. Delayed breeding can lead to over-conditioned heifers, which are harder to breed.

13.3.5 Target first lactation milk yields

The first lactation yield of heifers can be a useful guide as to how well they are grown up to the point of calving. Although their absolute milk yields can vary enormously with feeding management while milking, their milk yields relative to those of their herd mates is a useful criterion of heifer management. This value is determined by comparing the average full lactation milk yield of first lactation heifers with the average of the mature cows in the herd. Over the last 30 yr, this value has increased in Australian herd-tested farms from 65–70% to 80–85%. Friesian heifers have been grown out to produce 90% of the daily milk yields of their mature herd mates (producing 10 000 L/lactation) on Israeli feedlot farms.

If this value is 80% or less, heifer-rearing practices should be reviewed to establish if they are contributing to poor heifer production. Therefore, in a milking herd in which mature cows produce, say, 6000 L/lactation, heifers should be yielding at least 4800 L, while a mature cow average of 4000 L would correspond to heifers producing 3200 L/lactation.

13.3.6 Age of teeth eruption

It is easy to estimate the approximate age of a heifer by inspecting the state of her teeth. A calf may be born without teeth, with the temporary cheek teeth erupting within a few days and the temporary incisor teeth within 2 weeks.

The age at which the pairs of permanent incisor teeth erupt is as follows:

- first incisor teeth: 18–24 months
- second incisor teeth: 24–30 months
- third incisor teeth: 36 months
- fourth incisor teeth: 40–48 months.

The permanent cheek teeth erupt between 6 and 36 months, but are harder to identify than the incisor teeth. The age of eruption of permanent incisor teeth can vary with the feeding regime.

This is a very useful guide when objectively assessing the feeding management of young stock because poorly fed heifers may look healthy and relatively well grown, but if their first (or even second) incisor teeth have erupted they are likely to be much older than at first glance.

13.4 Feeding heifers to achieve target live weights

The formulation of rations to achieve target growth rates depends on available ingredients and their nutritive values. The expected economic benefits from applying the principles of target growth rates include:

- a shorter time before heifers generate farm income through milk production
- a faster rate of expanding the size of the milking herd

- increased farm income through higher heifer milk yields
- a reduced number of replacements required, therefore lower stocking capacity in the shed
- less forages required by growing heifers, hence more available for milking cows.

Once the growth targets have been determined, the rations must be formulated to provide sufficient dietary nutrients. More rapidly growing heifers with larger mature sizes require more protein in their diet, especially at younger ages. Such heifer feeding programs are likely to be more expensive. However, this approach to young stock management invariably leads to greater profits, particularly when well-grown heifers have the potential to produce more milk as mature cows.

Nutrient requirements for heifers depend on:

- expected mature live weight
- desired age at first calving; that is, acceptable age at first insemination
- current heifer live weight
- current age of heifer
- nutritive value of feeds offered.

There is almost always some check to growth at weaning, and this will be minimal if the animals are given gradual access to good-quality forages prior to weaning. During the wet season in the humid tropics, grazed pasture with limited concentrate supplements may be sufficient. However, in the dry or semi-arid tropics and during the dry season in the humid tropics, additional quality forages and concentrates are required to ensure good growth rates.

Before planning feeding strategies for growing heifers, it is important to set realistic target live weights for different ages. For heifers weighing 90 kg at 3 months to reach a target of 550 kg at calving as 2-yr-olds, they need to grow at 0.72 kg/day, compared with 0.57 kg/day if calving at 450 kg or 0.41 kg/day if calving at 350 kg. Average weight for ages and the energy (ME) intakes required to achieve these are presented in Table 13.6.

Table 13.6. Average weight for ages and requirements for metabolisable energy (ME) in heifers grown out to 350, 450 or 550 kg at 24 months of age

Age (months)	Live weight at 24 months of age and (average growth rate in kg/day)					
	350 kg (0.41 kg/day)		450 kg (0.57 kg/day)		550 kg (0.72 kg/day)	
	Live weight (kg)	ME intake (MJ/day)	Live weight (kg)	ME intake (MJ/day)	Live weight (kg)	ME intake (MJ/day)
3–6	127	29	141	34	155	39
6–9	165	33	194	41	221	49
9–12	202	38	246	48	287	59
12–15	240	43	298	55	353	68
15–18	277	49	349	64	418	80
18–21	314	51	402	66	484	82
21–24	350	55	450	70	550	85

Table 13.7. Dietary quality for heifers of different ages to grow at 0.7 kg/day

Age	3–6 months	6–12 months	>12 months
Energy (MJ/kg DM)	10.9	10.3	9.5
Crude protein (%)	16	12	12
Calcium (%)	0.52	0.41	0.29
Phosphorus (%)	0.31	0.30	0.23

Using an average ME level of 10 MJ/kg DM in the diet, the total DM intakes required to achieve the three growth rate targets presented in Table 13.6, can easily be calculated, by dividing the energy intakes by 10.

Heifers require a high-quality diet to grow at 0.7 kg/day. Table 13.7 presents the energy, protein, calcium and phosphorus concentrations of their diets to promote this rate of live weight gain. The limited rumen capacity of 3- to 6-month-old heifers means that they should be fed a ration containing as high an energy and protein concentration as that of milking cows.

The nutritive values of various feeds are presented are Chapter 10. A combination of good-quality tropical forages and concentrate supplements are required to produce a diet containing 9.5–10.5 MJ of ME/kg DM and 12% protein. Providing weaned heifers with a diet that is both short on daily DM intake as well as being low in energy content, is the major reason for sub-optimal heifer growth rates, which leads to light-weight heifers and delayed calving, hence reduced potential to produce milk and to get back in calf.

Feeding and grazing management should allow for continuous heifer growth throughout the first 2 yr. Uniform growth is not necessary, and may be impractical with fluctuating forage supplies. However, heifers should never lose weight or grow slowly for long periods during their first year, because they may not achieve their ultimate frame size and/or mating live weight by 15 months of age. Yearling heifers can show some compensatory gain in their second wet season following feed shortages the preceding dry season. Recommendations for grazing and feeding systems will vary with different regions. Rather than depend on 'recipes', farmers should use target growth rates to plan optimum feeding strategies. To achieve 550 kg by 2 yr of age, seasonal target growth rates can vary from, say, 0.5 to 1.0 kg/day.

When heifers mix with older cows, they increase their chances of picking up infections from, and hence developing immunities to, any diseases carried by the cows. These immunities can then be transferred to newborn calves via the heifers' colostrum. Heifers reared in complete isolation from cows are likely to become infected as they calve and come in contact with the milking herd for the first time. This coincides with the time when they should be in peak health to produce milk, get back in calf early and also overcome any stresses associated with their radical change in management. Earlier in their life, heifers were non-lactating animals continually at pasture whereas now they have become lactating animals with regular human contact twice each day.

If the milking herd has a history of Johne's disease or if there is a high chance that Johne's carrier cows have been introduced to the herd, then grazing options for young heifers are reduced. This, and other aspects of disease, are discussed in Chapter 11.

Agisting young stock off farm has much to commend it because it allows dairy farmers to use all available feed supplies to produce milk, while still having control over the disease status of the heifers and the genetic progress in the herd. However, farmers must be well aware of the supply and quality of pasture for their agisted stock, the responsibility for stock health while away from the farm and the security of the agistment area against theft and straying heifers, as well as neighbouring bulls. The proximity of the area and its cost are probably the major factors that need to be taken into account. Agistment works well, provided it is cost effective and heifer growth is monitored to ensure target weights are achieved.

Young stock should be handled frequently. When entering the milking herd, they must find their place in the social structure and this may take less time if they are used to human contact. For example, they should be run quietly through the milking shed a few times before calving to start to settle them into the milking routine. Grazing the heifers during their last months of pregnancy with the main herd of dry cows can accustom them to the competitive conditions with which they will have to cope during lactation. Hand feeding heifers for a few weeks before calving will provide extra feed to build up body condition, as well as get them used to being handled.

13.4.1 Grazing versus shedding of growing heifers

Much of Section 13.4 is based on the management of replacement heifers reared under grazing, whereas most SHD farmers in Asia rear their young stock under a continual shed environment. Compared with grazing, confinement creates specific problems such as:

- restricting the opportunity to seek comfort; for example, if only provided with cement floors
- creating problems of high humidity, which can be more detrimental than high temperature
- limiting opportunity for exercise, hence the need for routine hoof trimming
- increasing exposure to infectious diseases
- creating problems of heat detection for artificial insemination
- requiring greater efforts into sanitation
- magnifying problems of social dominance
- increasing capital investment.

There are two major reasons why SHD farmers do not have areas on their farm set aside for grazing: namely high land costs and cheap labour. There are other reasons and these include:

- the lack of farm forage areas
- inefficiencies of allowing stock to graze the forages rather than 'cut and carry' them to the stock
- better opportunities for climate control in a shed compared with the open air
- the better security of stock inside a shed compared with out in a paddock
- the difficulty of managing tropical forages to maintain quality while being grazed compared to a regular hand harvesting cycle.

Of all the classes of stock on a dairy farm, growing heifers would be the easiest to manage as grazing animals, because they do not require close attention at least twice daily, such as with milk-fed calves or milking cows. Because this is rare in tropical Asia, farmers can more easily plan feeding programs for growing heifers by providing adequate forages of acceptable quality rather than depend on their unknown ability to harvest sufficient nutrients themselves in a less-controlled grazing environment.

13.5 Contract heifer rearing

Well-reared heifer replacements represent the future for any dairy farm and regional dairy industry. As mentioned above, rearing heifers requires careful planning and adequate feed resources. As the cost of dairy land rises and the emphasis on milk production continues to drive profits, farmers with limited time and resources could consider contracting out the care of their replacement heifers. Such contracts could be with individual farmers, dairy cooperatives or even government support agencies. A good heifer rearer should, at the end of the contract, deliver animals in good condition, at an agreed weight for age and, if stipulated, pregnant. Specifications should be approved by both parties and documented in the contract.

Understanding the real cost of rearing heifers is the first step in developing an agreement with an independent rearing contractor. By assessing and placing a monetary value on the time and resources required to raise heifers on farm, the farmer can then determine the amount the service is worth to the dairy business when externally contracted.

The contract needs to document not only the agreed target weights but also the farmer's expectations on heifer management and conception. This written document should be agreed upon and signed by both dairy farmer and contractor. It should consider the following key points:

- target weights for age; these should be realistic for the breed type and the feed resources
- daily management: how often will the heifers be checked if grazing?
- routine animal health protocols, such as parasite protection, vaccination and drenching in accordance with district requirements
- veterinary attention: who pays the veterinary expenses?
- mating and pregnancy testing: whose responsibility is this?
- collection and delivery: who pays for transportation costs?
- the date for collection and delivery
- weighing procedure and frequency: specify intervals between weighing reports
- the payment schedule: what is the timing of payments? These should be staggered in line with animals reaching agreed interim target weights
- penalties for not achieving target weights: if the stock are not growing, the ability to remove them should be specified in the contract
- disease assurance program: an assurance that the rearing property is relatively disease free

- other stock: what other stock are on the rearing farm and will they be kept separate from the farmer's heifers?
- penalties for deaths and losses: it is recommended that a 3% allowance be given for stock deaths
- dispute resolution procedures; who arbitrates any disagreements?

When selecting a suitable contractor, dairy farmers could use the following checklist:

- stock skills and cattle handling facilities for weighing and providing for veterinary attention
- ability and facilities to manage an effective AI program, if desired
- resources to handle bulls such as fencing, yards and a separate bull paddock
- good external fencing with no neighbouring properties holding bulls
- knowledge of dairy heifers, their health and growth requirements, such as supplementary feed, mineral licks and good rotation through paddocks, for both grazing and cut and carry
- property in or near a common dairy area, so it is convenient, minimises transport costs and improves stocking rate flexibility during difficult seasons (that is, the stock can be easily returned home during feed shortages).

There may be local dairy farmers who wish to reduce their daily work load by changing from milking cows to rearing heifer replacements. Contract heifer rearing on owned or leased land can provide a solid cash flow business. To develop a successful heifer rearing business, the following points should be considered:

- Understand the costs and time required to rear dairy heifers.
- Talk to dairy farmers to understand the long-term risks and opportunities for contract heifer rearing in the district.
- Be in close proximity to dairy farmers.
- To decide on the size of the rearing operation, firstly assess the cost and availability of pastures and other forages and concentrate feeds, and then, secondly, take into account labour requirements and administration costs, such as any public liability insurance.
- Assess the risks in the paddocks where the heifers will be reared, if they are to be outdoors. Check fences, pasture, potential hazards and the location of neighbouring bulls to ensure heifers remain safe during their visit.
- Assess the potential pasture supplies and quality for the proposed stocking rate and the likely need for supplementary feeds.
- Once profitably is established, select the maximum number of heifers from a minimum number of farms.
- Encourage farmers to batch calve together, so the calves, being close in age, will be easier to handle, rear and mate, ensuring that record keeping and other management tasks can be kept to a minimum.
- Visit dairy farms from which calves originate and check the stock before transit to ensure they are in good condition and meet the contractual requirements.
- If taking young newly weaned heifer calves, ensure that all stock are fully weaned and unlikely to suffer any health or nutritional setbacks on arrival at their new farm.

- Only accept animals in good health after transit. Sick animals should be sent back to the dairy farmer.
- Prepare a detailed formal contract based on the key points above.

While the onus is on the dairy farmer to set the targets, flexibility and common sense should prevail with concessions made for unusual and unexpected circumstances. Long-term contracts are more likely to develop if the farmer's expectations are realistic and clearly communicated to the rearing contractor.

Contract heifer rearing can be undertaken by a commercial operator or, as is becoming increasingly common in tropical Asia, constitute one of the services provided by dairy cooperatives or even government livestock agencies. It is highly unlikely that any Asian country will be able to achieve realistic target national herd sizes without importing dairy heifers. This is mainly because of the very high pre- and post-weaning mortality rates referred to in Chapter 6, which can only be overcome through developing the skills discussed in this book.

A commercial contract calf rearer will have to become competent quickly to remain in business. The development of 'heifer farms' is a new initiative of governments and cooperative agencies, removing the calf and heifer rearing responsibilities from individual SHD farmers; the pregnant heifer is then returned to that farm just prior to calving down. Such heifer-rearing services provides the opportunity for bulk purchase (and routine nutritive testing) of feeds, specialist skills in animal health and production, and good rearing facilities. It can also allow for the importation and bulk rearing of young (say 6-month-old) dairy heifers, which would reduce their purchase and transport costs and give them more time to adjust to the tropical environment before mating and eventual pregnancy.

13.6 Using dairy stock for beef production

Although dairy cows need to produce a calf each year to continue milking, dairy farmers only require 20–30% of these calves to replace those cows that die or are culled from the milking herd, because the average productive life span for milking cows is four to five lactations.

Artificial rearing of replacement heifer calves is a specialist job that dairy farmers generally have to undertake. Because it is the most expensive period in an animal's life, calves reared for dairy beef would need to be grown out for slaughter at much older ages to dilute these high initial feeding costs. Unless these finished animals realise reasonable returns, dairy beef is unlikely to be profitable.

No matter what the system of beef production, the maintenance of breeding stock and the production of their offspring to replace those slaughtered for human consumption is a major part of the total feed inputs. With beef producers, these must be included in their total production costs. However, with dairy beef producers, these are 'paid for' by dairy farmers. Because dairy calves are by-products of the milk industry, dairy beef farmers should then have lower production costs than beef farmers. Their ultimate lower carcass returns can allow for this, but they have rarely been sufficient for dairy beef to have a long-term viable future.

Growing out bull calves for slaughter at 18 months or more is frequently a major generator of income for dairy farmers. These stock can be kept entire and grown out as bulls, or they can be castrated at an early age and grown out as steers. Bulls grow faster than steers, but, when they reach puberty, they can become a nuisance with milking cows on heat. The same principles of feeding management for dairy heifer replacements should also be used for growing out dairy beef bulls or steers. Target growth rates should be at least 0.5 kg/day. There may be additional recommendations with regards to vaccinations and other health management issues, and such information should be sought from local animal health specialists.

Purebred or crossbred Friesians are preferred over Jerseys or local breeds and generally return more money per kg live weight when sold for dairy beef, so the breed type of the bull calf should be considered when deciding on its future fate on the farm. On the whole, milk generates more income per kg fresh grass grown on the farm than beef does, so it is important to 'do the sums' on dairy beef production when farm stocking rates are high. Acceptable growth rates are unlikely if the major part of the ration is very low-quality forages, such as rice straw. As with milking cows, green grass is the best forage to feed dairy beef animals.

It is important to plan the finishing (or fattening) strategy for dairy beef. Target live weights should be used in any feeding program, so using a chest girth tape to estimate live weight is worth considering. Friesians are later maturing than other dairy beef animals, meaning that a higher level of concentrate feeding will be required to produce a suitable sale animal: namely one with a good degree of finish (or some cutaneous fat cover over the ribs).

14

Mating and calving management of dairy heifers

This chapter discusses the optimum management of dairy heifers during mating and their first calving.

The main points in this chapter

- The key decisions when mating dairy heifers are: what age and weight should they be and what type of mating (natural or artificial) should be used?
- Mating heifers at too low live weights will lead to reduced fertility and more calving difficulties. Mating pure or crossbred Friesians weighing less than 260 kg will lead to more calving difficulties, as will mating Zebu dairy-type or Jersey heifers at less than 200 kg.
- 'Catch up' feeding after mating often results in heavier calves at birth, over-conditioned heifers and more calving problems, with little improvement in milk yield during the first lactation.
- A herd of 100 cows generally requires the rearing of 20 to 30 heifer replacements each year. Bulls (or semen) must be selected on 'ease of calving' to reduce potential calving difficulties.
- Natural mating of heifers is easier than artificial insemination (AI) and running heifers with a bull for 9 weeks can lead to good pregnancy rates, provided the heifers are actually cycling.
- AI will provide greater scope for selection and genetic improvement in the dairy herd, but it requires a higher level of skill in heat detection and insemination. Natural mating is often required after AI to further increase conception rates.
- Heat detection aids can be used, such as tail paint, pressure-sensitive heat mount detectors, vasectomised bulls or hormone-treated steers. Heat synchronisation also reduces the time required for heat detection.
- Sexed semen can produce 90% heifer calves in a well-managed breeding program. However, in addition to its considerably higher cost, conception rates are lower than with conventional semen (40% versus 50% with virgin heifers).

- It is important that calving heifers are provided with a good, clean calving area to minimise injuries to the uterus and reduce infections in newborn calves.
- The newborn calves should be permanently identified, their navel cords disinfected with iodine and they should be assisted to suckle their dams or be provided with adequate good-quality colostrum.

This chapter covers two key areas of post-weaning dairy heifer management: namely mating and calving down the first calf. Figure 14.1 provides a visual summary.

14.1 Mating replacement heifers

When deciding on the most appropriate mating management for their replacement heifers, farmers need to ask a series of questions, such as:

- What age should heifers be at mating?
- How big or heavy should heifers be at mating?
- What type of mating, natural or artificial, should be used?

A normal pregnancy takes 282 days from conception to calving. If heifers are to calve at 24 months of age, they should be mated at 15 months. The only time when heifers should calve older than 24 months is when their growth has been restricted and they are not sufficiently developed to mate at 15 months.

Most changes in skeletal size occur before puberty. The onset of puberty – that is, the commencement of cycling – is related to live weight rather than age. A delay in puberty

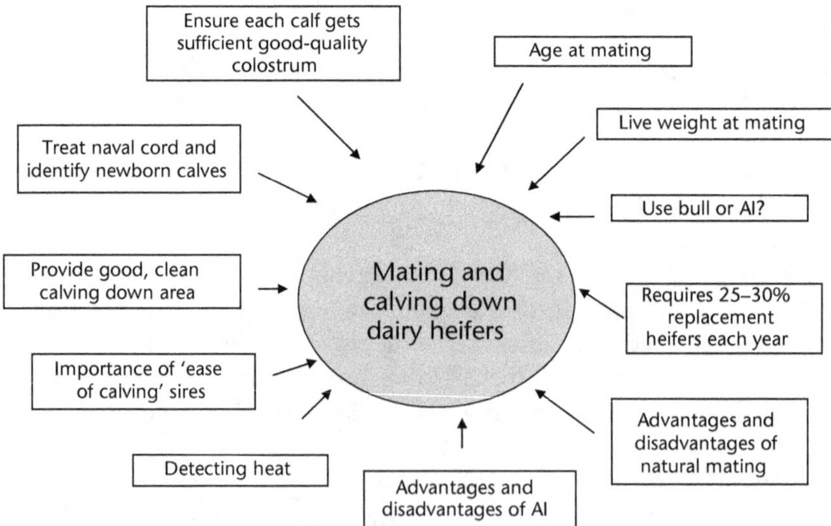

Figure 14.1. The key factors to consider when planning mating and calving down of dairy heifers

means an increased chance of a delayed first conception, which can disrupt future calving patterns.

All heifers should reach minimum live weight before mating, because lighter animals have lower conception rates and more calving difficulties. Mating pure or crossbred Friesians weighing less than 260 kg will lead to more calving difficulties, as will mating Zebu dairy-type or Jersey heifers at less than 200 kg. 'Catch up' feeding after mating often results in heavier calves at birth, over-conditioned heifers and more calving problems, with little improvement in milk yield during the first lactation.

Data from tropical breeds are limited, but it is likely that the normal variation in the gestation period is between 275 and 287 days, as in temperate zones. The normal variation for heats, 18 to 24 days, is also similar to temperate regions. There is, however, a major difference in the duration of the heat. In temperate zones, it averages 18 hr, ranging from 6 to 30 hr. In both temperate and tropical breeds in the tropics, the average is much shorter and this short period often occurs at night, when 'silent' heats are more common. This complicates breeding management, with often low fertility unless special managerial measures are employed. This is particularly so for grade Friesian dairy herds.

AI should be used only when the support infrastructure is available and herd management is good. Even with AI, a back up bull should be available to breed heifers that miss the normal AI program. Heat detection may be improved with vasectomised bulls or using heat mount detectors, located on the back of cows, that burst when she allows another animal to mount her.

Dairy cows usually come into heat 30 to 60 days after calving. A percentage of heifers and cows are always sterile and in temperate zone this varies from 3% to 5% in cows up to 10 yr of age; the incidence is higher in older cows. Culling for sterility is essential, while short heat periods and silent heats can be reduced by aggressive culling. Once the cows conceive, 2–3% may lose their calf due to abortions.

Batch calving, or calving down cows at particular times of the year can lead to easier feeding, such as in the early wet season when forage supplies and quality are at their best, or be part of breeding management, such as not trying to mate cows in the hottest season. However, because most small holders are seeking a regular milk income, ideally they should be calving cows down throughout the year.

Heifers should be gaining weight for at least 60 days before mating, be in good body condition and be seen to be cycling. Heifers in poor condition and losing or only maintaining weight will have reduced fertility.

Calving difficulties (or dystocia) are common concerns with heifers and the major causes are:

- using a sire which produces too large a calf for the size of the dam
- poor feeding management both before and after mating
- problems with presentation of the calf at birth
- heifer- or cow-related problems with the anatomy of the birth canal.

With heifers, the first and second problems are the most common causes of dystocia, whereas with older cows, it is the third and fourth problems.

14.2 Natural mating

Natural mating is using bulls to service the heifers. The choice of bull depends on many factors, such as:

- the heifer's age
- the heifer's size or live weight
- the heifer's stage of development
- farmer's requirement for extra heifer replacements to increase herd size or achieve desired culling rates
- availability and cost of bulls.

Using beef sires over replacement heifers is a common practice on many dairy farms. The argument in favour of using beef bulls has been that small poorly developed dairy heifers are not big enough at 15 months to be serviced by dairy bulls (except in the case of Friesian heifers, other than by Jersey bulls). Problems can occur with physical damage when small heifers are mated with large bulls, because calving difficulties frequently occur with large calves being born to small heifers. However, heifers that have achieved minimum target live weight are generally capable of being successfully mated and calved to the larger dairy breeds. With the recent influx of large-framed overseas genetics into what were previously considered small and safe, easy calving beef breeds, such as Angus, many producers now experience calving difficulties through mating their smaller heifers with these beef bulls. If newborn 'bobby calf' prices are low, the value of a beef cross calf is minimal compared with the potential cost of a dystocia and the resultant loss of milk production.

The advantages of natural mating over AI include:

- There is no cost of semen or equipment needed for semen storage and insemination.
- Natural mating requires less labour with a lower level of skill than AI.
- If the bull is run with the herd for periods less than 9 weeks, good pregnancy rates can be achieved, because the submission rate is extremely high provided the heifers are actually cycling.
- No heat detection is required.

Some of the disadvantages of natural mating are:

- The genetic gain with natural mating is less than with AI.
- Capital invested in bulls could be used for other productive investments.
- Feed eaten by bulls could be used to feed cows to produce milk.
- Bulls require more secure fencing.
- Bulls are more likely to damage fences and other equipment.
- Running bulls with heifers on agistment may not be acceptable to the land owner or they might restrict the selection of the desired bull for agisted heifers.
- There is a risk of severe mating failure through infertile or lame bulls or bulls with low libido.
- Natural mating introduces the risk of venereal disease, such as vibriosis.

- Natural mating also increases the risk of disease introduction to otherwise closed herds, such as Johne's disease.
- There is an increased risk of injury to people when managing bulls.

14.2.1 Bull management for natural mating

Despite attention to achieving heavier mating live weight and the publication of calving ease values for AI bulls, many farmers still find that Friesian heifers mated to Friesian bulls leads to too many difficult calvings, compared with AI or natural mating with Jerseys. They may still prefer to use Jerseys for mating their Friesian heifers, either for ease of calving and/or to produce herd replacements. This can be a practical option because herd recording data show that Jersey × Friesian cows can be very competitive in milk production, and may have superior feed efficiency, after taking into account their lower live weights.

Sufficient bulls should be run with each heifer group. Too few bulls is a common problem, resulting in low pregnancy rates. The recommended rate is one bull for every 30 empty heifers plus one spare bull. The number can be reduced as more heifers become pregnant. For example, after a single AI, about 50% of the heifers are likely to be pregnant and only half the bulls will be needed compared with natural mating alone. However, if the heifers have been synchronised, their first heat during natural mating will be compressed so that twice as many will be on heat per day, and more bulls should be used for this first cycle.

To ensure good results with natural mating, bulls should be physically examined for reproductive abnormalities and observed to ensure they can mount, serve and ejaculate normally. They should be vaccinated for vibriosis. Their feet need to be inspected 6 weeks before mating and trimmed if they are overgrown. Lame or overweight bulls will not be able to serve their full quota of heifers, thus should be avoided, replaced or augmented with other fit bulls. The health of bulls should also be tested prior to introduction to the herd, or at least with Johnes disease, be accompanied with a vendor's declaration of a disease-free status.

Any heifer not in calf after a 3-month mating period should be culled because she is likely to have permanent low fertility: a trait worth culling to avoid.

14.3 Artificial insemination

Farmers can achieve the greatest genetic gain in their herd by mating maiden heifers with AI, using semen from 'proven' dairy bulls selected for milk, fat and protein, and a rating for ease of calving. These advantages arise from:

- production benefits from their progeny
- reducing the generation interval
- increasing the number of calves of higher genetic merit available as herd replacements.

Other advantages of AI over natural mating include:

- reducing the number of bulls required on the farm
- having a shorter mating program, particularly if using oestrus synchronisation, thus resulting in a shorter calving period and less time spent on calf rearing
- reducing the risk of disease introduction from purchased bulls
- reducing the risk of reproductive failure through bull infertility, injury or low libido
- increased handling of heifers for AI can improve their temperament after calving.

The disadvantages of AI include:

- added costs for semen purchase and storage
- added cost of heat synchronisation, if used
- increased requirement for skilled labour for heat detection and insemination
- heifers have shorter cycles than mature cows, making heat detection more difficult
- submission and conception rates can be lower, although less variable
- heifer groups often calve over a slightly longer calving period
- if heifers are away on agistment, AI programs are less convenient
- natural mating is often required after AI to ensure adequate conceptions.

It can take up to six straws of semen to generate one replacement heifer in the herd, so for a 50 milking cow herd with 25% replacement rate each year, this requires a total of 75 straws. This is based on the following assumptions:

- 50% conception rate per straw
- 10% loss of cow prior to calving (that is, pregnant cow that either dies or is culled)
- 50% female calves
- 5% loss of heifers pre-weaning
- 5% loss of heifers prior to joining
- 90% heifer conception rate to natural mating
- 3% loss prior to calving
- 3% loss within 30 days post-calving.

The traditional management system of naturally mating with terminal sires leads to a 5 yr interval between the birth of replacement heifers to the entry of their progeny into the milking herd. If heifers are mated to AI at 15 months, their first calves will enter the herd when they are 4 yr old.

A herd of 100 cows requires 20 to 30 heifer replacements each year. If beef bulls are used on heifers, these replacements must be bred from nearly all the early cycling cows, allowing little chance of selecting the better cows from which to breed future replacements. Using AI over the heifers also provides greater scope for selection and genetic improvement of the herd.

Ease of calving depends on the condition of the dam and calf factors such as gestation length, calf weight and presentation at birth. A bull rated highly for calving ease can still produce calves that have the potential to grow into large-framed cows, even though they have lower birth weights. When using AI, it is important to select easy calving sires. These should have 'calving ease' values of 2–3%; that is, only 2–3% of the calves born to that sire in mature cows will present calving difficulties. It is important to note that 'calving ease' values are determined from mature cows, meaning that they may underestimate difficult calvings in heifers.

14.3.1 Managing artificial insemination

Heat detection

Heifer heats are much less obvious and of shorter duration than those of older cows. An 8-hr heat is not uncommon, so successful heat detection in an AI program requires the input of time from the herd manager.

The incidence of cows showing signs of heat vary throughout the day. Recent US studies with dairy cows (Anon 2010) found that 22% of the cows showed heat between 0600 and 1200 hr, 10% between 1200 and 1800 hr, 25% between 1800 and 2400 hr while the highest incidence, 43% of the cows, showed heat between 2400 and 0600 hr the next day.

Heifers must be observed for at least 20 min, three times each day, to detect heat. Because most producers cannot allocate this much time to AI of their heifers, they often use aids or drugs to assist with heat detection.

Heat-detection aids

These include tail paint, pressure sensitive heat mount detectors (KAMAR®) and vasectomised bulls or hormone-treated steers with chin ball markers. Recent technology is also now available, either using electronic pressure-sensitive pads to monitor heat detection (Heat Watch®) or pedometers to measure the number of steps taken over 24 hr. No single heat detection system can be recommended for every herd because factors vary between herds and most systems can work well. The choice of heat-detection aids depends on many management factors and whatever the system chosen, it will have to be managed well to ensure good results.

Tail paint is applied in a strip about 20 cm long and 10 cm wide, centred on the highest point of the tailhead where it will be disturbed when the heifer is mounted. A heifer is classified as on heat when the tail paint is disturbed. Any exterior house paint can be used and bright colours are easy to see. A different colour paint should be used after insemination to help detect heifers that have not conceived, so they can be mated again using AI or conceive later to a natural mating.

There are many different ways to use tail paint and most will work if carefully managed. One system that works well is to apply fresh tail paint every 3 days and observe heifers very carefully in the paddock or in a race twice per day, then inseminate any with tail paint that is not clean and shiny, but was clean and shiny at the last observation.

KAMAR® heat detectors are glued onto the highest point of the tailhead. When rubbed firmly, following mounting from another animal, a capsule of dye breaks, indicating that this heifer is probably on heat. Both KAMAR® heat detectors and tail paint can give 'false readings', and so should only be used as aids to heat detection. Other signs of heat should be observed at the same time to confirm oestrus, or poor pregnancy rates may result.

Heat synchronisation

Heat synchronisation reduces the time required for heat detection. A group of synchronised heifers can be managed more easily and precisely because they are all at the same stage of pregnancy. Synchronisation also concentrates the resulting calving over a shorter period, reducing the time needed to oversee calving. AI-bred calves from synchronised heifers are born closer together, reducing management problems during calf rearing. Optimum results will only be achieved under ideal conditions, such as during fine weather, with access to good yards and handling facilities and when heifers are in good body condition. If conditions are not ideal, it may be advisable to use natural mating rather than AI.

In seasonal calving herds, a heifer synchronisation program can advance the average calving date by up to 10 days. This is partially offset by reduced submission rates. If heifers are mated, say, 10 days before the cows, then their earlier calving date will give them ten extra days prior to the start of the next mating. This is important for first calving heifers because they take longer to recover from the stresses of calving and commence cycling.

Cost is the main disadvantage of synchronisation. Careful planning of the synchronisation and AI are needed, while additional skilled labour is required to successfully manage the concentrated mating and calving periods. It is advisable for 'do it yourself' inseminators to employ a trained (A class) inseminator for their synchronisation programs, unless they have been able to refresh their technique and practice on some cull cows just prior to the program.

As a general principle, inseminations should be when heat is detected. Fixed time inseminations, also known as 'blanket insemination' (that is, without heat detection), give variable results, and should only be considered if conditions are ideal. Some groups of heifers achieve more than 70% pregnancy to AI in a single, synchronised cycle, while other groups may only achieve 20% pregnancy. Low AI conception rates can be compensated for by natural mating after insemination to get most heifers to calve in the desired calving period. Some farmers are willing to accept variable AI conception rates because there is less work involved when heat detection is not used.

Near-term calving induction has been used successfully following synchronised AI programs, with heifers induced to calve 7–10 days before full term. They generally calve about 36 hr later, ensuring smaller calves with less calving difficulty. This enables any required assistance to be given promptly. It also reduces night-time calvings and can eliminate weekend calvings.

14.3.2 Examples of heat synchronisation

Heifers can be synchronised in many ways to compress the normal 21-day cycle. Generally speaking, the shorter the cycle achieved, the more costly the synchronisation treatment. Some examples of synchronisation methods with approximate drug costs (in Australian dollars) and spread from first to last AI are:

1. Heat detection and insemination for five days followed by a prostaglandin injection, and a further five days of heat detection and insemination. A bull is usually run with the heifer group for a further six weeks. This is a low cost option that reduces the heat detection and insemination period from 21 to 10 days and still gives each heifer

one opportunity for AI. It is also the most time consuming synchronisation option. The prostaglandin costs about A$4 per heifer. This will typically result in 60% of the heifers calving to AI.

2. Using a progesterone implant, such as a CIDR (controlled internal drug release) device. These intravaginal implants are inserted for 7 days with a dose of oestradiol at the start. A prostaglandin injection is given at the time of implant removal. This is followed by 4 days of heat detection and insemination, with an oestradiol injection 2 days after the device is removed. Again, a bull is usually run with the group for a further 6 weeks. This compresses the heat detection into 4 days from the normal 21-day cycle but costs about A$20 per heifer, with about 60% of the heifers calving to AI.
3. Using an intravaginal progesterone implant for 7 days with a dose of GnRH at the start. A prostaglandin injection is used at the removal of the device and a second GnRH dose 2 days after removal. All heifers with detected heat are inseminated, then, after 56 days, all other heifers (previously not detected on heat) are blanket inseminated. A bull is then introduced for 6 weeks. This removes the need for heat detection completely but costs more than A$20 per heifer. The results vary widely, with blanket insemination, but a long-term average of about 50% of heifers pregnant to AI can be expected over a number of years.

The programs require multiple yardings, so heifers must be quiet and used to being handled. Hygiene, correct semen handling and good AI technique are also critical. Many of these drugs can only be supplied by veterinarians, whose responsibility is to prescribe the program and drugs used.

Some treated heifers may produce an unpleasant discharge at CIDR removal, although this is unlikely to adversely affect fertility. A few heifers may also suffer damage from CIDRs due to pressure on the vaginal wall. The biggest single cause of failure in synchronisation programs is low mating live weight, hence non-cycling heifers.

14.3.3 Sexed semen

Sexed semen is a relatively new commercial innovation in advanced reproductive technology, but it is still being evaluated for its long-term benefits, particularly on farms where breeding management is sub-optimal. This semen has a very high proportion of female sperm cells, which can result in up to 90% heifer calves when used in a well-managed breeding program. Another of its advantages is a reduction in calving difficulties when used to mate virgin heifers (Millar 2010), because of the lower birth weight of heifer calves. Some Asian countries are producing their own sexed semen using less advanced techniques and only achieving 70% heifer calves.

However, its biggest disadvantage is a lower conception rate than with conventional semen. If 50% conception is expected using AI with virgin heifers, a similar group of heifers mated with sexed semen would be expected to result in 40% conception. So, in addition to its considerable higher cost (A$60 per straw versus, say, A$20 per straw in Australia in May 2010), there will be fewer conceptions than with conventional semen. Accurate heat detection is critical when using sexed semen and it is vital that animals be inseminated in standing heat. Proper semen handling and AI technique is also critical. One other disadvantage is the limited range of bulls from which sexed semen is available.

In fact, with any imported, exotic semen one must query whether the bulls from which the semen is collected are the most suitable for the tropical SHD production systems for which they are purchased. The interaction between genotype and environment is discussed in my previous books (Moran 2005, 2012).

14.3.4 Targets for heifer mating programs

The Dairy Calf and Heifer Association in the US published a series of targets for heifer mating programs for Friesians (DCHA 2010), which can be used as guidelines for tropical dairy systems. These are:

- Begin mating at 13–15 months of age with heifers weighing 375–410 kg, with hip height of at least 127 cm and wither height of at least 122 cm.
- Strive to achieve 70% first service conception, or 7–12% less with sexed semen.
- Inseminate at least 80% of the heifers within the first 21 days of mating.
- 85% of the heifers should be pregnant within three heat cycles.
- Test all heifers to confirm pregnancies (some animals – typically 3% – will abort after pregnancy testing).

14.4 Calving down replacement heifers

Dystocia or calving difficulties can contribute to death loss at calving. Selecting light weight bulls is the best management strategy to reduce the incidence of dystocia. Measuring the pelvic area of yearling heifers prior to breeding them is a good way to check whether heifers can handle calving. Heifer development programs should aim for heifers to reach 65% of their mature weight at breeding and 85% at time of first calving.

It is a misconception that reducing the feeding levels prior to calving will reduce dystocia. This will reduce calf birth weight, but not dystocia. It may, in fact, reduce calf survival and increase calf scours. Animals should not be too fat or too thin at calving, to avoid calving difficulties, metabolic disorders and potential subsequent culling.

Parturition is the term used for giving birth to the young calf. The calf can be expected 9 months and 9 days (40 weeks) after conception. The udder will start to swell 2–3 weeks before calving. Within 12–24 hr of parturition, the heifer requires a comfortable clean place where she can lie down easily. She should be cleaned before entering this area to avoid introducing unnecessary contamination. Farmers should also practice good hygiene by washing their hands and the rectal/vaginal area of the cows.

A good calving area is important for:

- the heifer, to avoid injuries and infections of the uterus and to stay there 24–48 hr post-calving
- the calf, to avoid infections of the navel passed on from other stock that either are, or have been, in the shed

- the farmer, to have enough space to assist with the calving. It should be at least 3 × 4 m in size.

The calving area should be:

- isolated from other stock
- have good light, ventilation and protection against any winter draughts
- be easy to clean and disinfect
- provide easy access to fresh, cool and clean water
- covered with a generous layer of straw, to minimise the heifer's problems with slipping
- ideally, not be used for any other purpose, particularly as a hospital pen.

14.4.1 The actual birth process

Cows undergo a series of physical changes that indicate the start of calving. Several weeks or days before parturition, the udder fills with colostrum, the vulva becomes swollen and the pelvic ligaments relax, causing a sunken appearance around the tailhead. Mucus discharged from the vulva is no longer thick or opaque, but becomes flowing and clear, similar to the mucus observed during oestrus. The actual birth process progresses through three continuous stages: 1. preparation for delivery; 2. expulsion of the foetus; and 3. expulsion of the foetal membranes. The events of each stage are described in Table 14.1.

14.4.2 Monitoring the actual birth process

It is important to observe calving cows and heifers every 2 hr and to monitor the degree of difficulty and the assistance required during calving. This can be done using a calving score based on the benchmarks in Table 14.2.

In many cases, extended labour as well as lack of adequate lubrication in the birth canal increase the likelihood of calving problems. It is important to:

- observe the calving heifer (or cow) for labour signs
- examine the dam that is not making normal progress once in labour
- intervene soon enough to reduce the help required later for a successful birth. However, it is important to give the calving heifer sufficient time (up to 2 hr) before intervening.

14.4.3 Assisting the birth process

Assistance to calving heifers and cows should be provided when any of the following occur:

- When heifers have not calved within 5–6 hr after the first sign of abdominal straining.
- When cows have not calved within 3–4 hr after the first sign of abdominal straining.
- When calving has not occurred within 3–4 hr after the membranes have ruptured.
- If delivery has commenced, the calf's legs or head are just visible externally and it is obvious that the calf's presentation is abnormal.

Table 14.1. Events of a normal delivery

Stage	Events	Duration	Action
1	Relaxation and dilation of cervix	Cows: 3–6 hr Heifers: 3–24 hr	Observe frequently from a distance
	Contractions last 15–30 seconds, 15 min apart		
	Cow acts restless, seeks solitude, may look at flanks, stamp feet, raise tail or arch back		Examine animal if no progress after 5–8 hr
	Weak straining		
	Ends when foetal membranes become visible		
2	Begins when water bag breaks	Cows: 0.5–1 hr Heifers: 1–4 hr	Continue observations every 30 min
	Legs and head push through cervix into vagina		
	Cow acts restless, stands and lies down		Examine animal if no progress after 1 hr
	Strong straining		
	Contractions last 10–15 seconds, every 2–3 min		
	When feet enter vagina, second membrane breaks, releasing thick lubricating fluid		Examine heifer if no progress after 2 hr
	Cow rests after head and shoulders delivered Remains of body passes quickly		Examine for malpresentation, such as breech delivery
			Examine if cow showing any signs of calving paralysis or distress
3	Continued contractions	Cows: 2–8 hr Heifers: 2–8 hr	Considered retained placenta if not passed within 12 hr
	Rapid decrease in size of uterus (involution)		Do not attempt to remove placenta manually
	Attachment between uterus and placenta relaxes and placenta separates		Do not insert uterine bolus as it may prolong passage
	Ends with passage of foetal membranes (after birth)		

Table 14.2. Benchmarks for calving scores

Calving score	Birth process
0	No hand touches the calf
1	Hand touches the calf but no rope is used
2	Rope is used: gentle pull
3	Rope is used: hard pull
4	Caesarean section, requiring anaesthetising the heifer for a full operation

- If delivery has commenced, the calf's legs or head are just visible externally and the calf is not delivered within 30 min.

If you think that a cow may have calved (for example, she may have placenta hanging from the vulva) but you have not found the calf, it is best to perform a vaginal examination to ensure that she has in fact calved.

When assisting a calving cow, cleanliness is essential to prevent the introduction of infection. Use plenty of warm water containing disinfectant to clean the back end of the cow and thoroughly wash your arms with clean, fresh warm water containing disinfectant. Lubricate your arms with obstetrical lubricant, always using plenty of lubrication to stop the calving passage from drying out.

Investigate the presentation of the calf in a systematic manner and be gentle. Check that the cervix is dilated by inserting your lubricated arm into the vagina. If you feel a band, or there is a ridge between the uterus and the vagina, then do not attempt to pull the calf. If two hind limbs can be felt, check that the hocks and the calf's tail can be positively identified. If the calf's head cannot be felt, do not assume that the legs are the hind legs, they may be the front legs and the calf's head may be twisted back. Traction is often necessary to deliver an oversized calf and in cases where active straining has stopped. Great force should not be used when traction is applied and never use vehicles to apply traction.

Do not attempt to deliver the calf unless it is:

- upright with both forelegs first and head lying over its knees
- upright, backwards with two hind legs and tail presented.

If the head and legs are not in the normal positions, they can often be gently moved into place. Never pull the calf until it is positioned correctly.

If the calf is coming forwards, traction should be applied at three points using:

- a cord or calving chain with a running noose applied to each foreleg above the fetlock
- a third cord with a running noose placed behind both ears and between the jaws (or eye hooks, as in this case) – but do not put a noose on the lower jaw because the jaw will break.

The direction of pull is critical and the calf is delivered in an arc. Once the head and feet are through the vulva, the pull should be directed towards the cow's hocks. If the calf's hips become stuck in the pelvis, do not pull, but rotate the calf's hips 14 degrees from the vertical then pull in a direction towards the cow's hocks. Always check to make sure that there is not a second (or third) calf in the uterus.

Hygiene, antibiotics and oxytocin will reduce the risk of uterine infection following an assisted calving.

Call the veterinarian if you cannot:

- work out what is wrong
- correct the position of the calf within 5–10 min
- pull the calf with three people.

Table 14.3. Calf mortality by dystocia category

	Percentage of calvings		Calf mortality within 48 hour (%)	
Calving difficulty	**Heifers**	**Cows**	**Heifers**	**Cows**
Unassisted	45	79	8	6
Easy pull	30	15	10	8
Hard pull	14	3	35	24
Calf jack	7	1	55	66
Veterinarian	4	1	48	65

Prolonged assistance can severely damage the birth canal and result in the death of the cow and/or calf. A safe delivery increases the cow's chances of breeding next season and increases the chance of a live calf.

Farmers should aim to have less than 7% of their cows and less than 10% of their heifers calve down with severe assistance (scores of 3 or 4). The chances of these calves dying is over six times greater than for calves requiring no or little assistance (scores 0 and 1). The major causes for dystocia in heifers are either selecting a sire that produces heavy calves or poor feeding management during pregnancy.

Table 14.3 presents a survey of calf mortalities related to different degrees of calving difficulties undertaken in the US over 20 yr ago (Heinrichs and Swartz 1990). The categories for calving difficulties do not correspond directly to those described in Table 14.2. This clearly shows a greater likelihood of calving difficulties in heifers than in cows and the associated very high calf mortalities (50–60%) among these calves.

In the US, 2% of calves are born dead. The live birth rate should also be monitored. Live birth rate is defined as the percentage of calves that are alive and survive for at least 24 hr. Targets are 97% for cows and 92% for heifers.

The long-term impacts of calving difficulties can be quite dramatic for both the cow and the calf. Calving heifers receiving assistance take longer for their first service in the future, experience more services per conception, and hence have a longer calving interval (by up to 28 days), compared with heifers not requiring assistance. In addition, calves experiencing calving difficulties produce less milk in their first lactation.

14.4.4 Caring for the newborn calf

Every time a heifer calves down, she contaminates the calving area with 15–20 L of amniotic fluids (that surround the developing calf foetus) mixed with faeces. Dry bedding on a concrete floor can provide an effective physical barrier between the newborn calf and the pathogens. Immediately after calving, the calf should be provided with clean bedding where the dam may lick the calf. It is important to minimise the possibility of pathogens entering the calf via the two major routes: the mouth and the navel.

It is essential to make sure the calf is breathing by clearing the nose and mouth. If breathing is slow to occur and sticking straw up the nose does not initiate breathing, cold water could be poured over the calf's head, especially into its ear canal, and its nose could

Calves should be removed from their dams after being cleaned.

be tickled. It could even be held upside down. The calf should be placed in a 'dog sitting' position to encourage inhaling through the lungs. The navel cord should be disinfected with 7% iodine and checked for infection (pain or swelling); this should be repeated 24 hr later. Death rates of calves without navel disinfection are 10–15% higher than those that have been disinfected. Teat dip should not be used to disinfect calf navels because it contains less iodine and does not contain alcohol, which dries the navel.

If desired, the calf could be assisted to suckle from its mother soon after birth to ensure it drinks some colostrum, which provides protection against various infections. Assisting suckling should at least allow the calf to be offered a relatively clean teat. If there is no natural suckling, the heifer should be hand milked with the first feeding of colostrum given using a bottle and teat or a stomach feeder.

The longer the calf spends with her dam, the greater the chance of exposure to pathogens, and therefore contamination. But how long is too long? The dam should be allowed to lick the newborn calf because this also stimulates breathing and blood circulation. Once the calf stands and is ready to walk, its chances of picking up bacteria from the dam's hair coat, dirty teats or from falling into manure greatly increase. Therefore it is best to help the calf find a clean teat for its first drink and then separate it from its dam. This may seem rather cruel, but producing a clean and healthy calf should be the primary objective from every calving.

The three golden rules to ensure newborn heifer calves get a healthy start to life are:

1. Get her up.
2. Get her dry; this may be better achieved with a dry towel and vigorous rubbing to get a dry fluffy coat, rather than depending on the dam to lick her dry.
3. Get her fed.

Calves should be suitably identified (permanently) soon after birth, with only healthy calves kept for rearing. When the cow has twins, there can be problems with possible fertility if the calves are of different genders. The heifer calf, called a freemartin, may not be fertile, so cannot be used as a dairy herd replacement.

Calves resulting from difficult births require close attention over their first few weeks of life. These calves have higher sickness and death rates. They should be individually identified, say, with coloured string around their necks or a mark on their head, and watched more carefully for early signs of scours or pneumonia.

By 30 min after birth, the gums and nose should have changed to pink (from bluish-grey at birth). Note any abnormalities such as:

- swelling of the head and tongue
- dopey behaviour
- cloudy or bloody eyes
- head held back, arched back or puffy abdomen
- blood from any opening
- laboured breathing
- broken ribs.

Monitor these calves closely. By one hour after birth, the calf should be able to stand and have a normal body temperature (near 38.6°C). When the calf is born, its body temperature is 39.4–40.0°C. It drops to about 38.3°C within 30 min, but should stabilise by 1 hr. It is important that the calf is placed in its permanent calf housing by this time to allow it to adjust in that environment. If the calf's temperature is less than 37.8°C, it should be provided with an artificial heat source, such as a heat lamp or blanket, to be warmed.

In summary, for a normal delivery, a newborn calf should exhibit the following signs:

- the calf tries to lift its head within minutes
- it rolls up on its sternum within 5 min
- it attempts to stand within 15 min
- it is standing within 1 hr
- its temperature, while higher at birth, stabilises to 38.3–38.9°C by one hour after birth
- it suckles within 2 hr
- its respiration rate is 50–75 breaths/min
- its heart rate is 100–150 beats/min.

If most calves do not exhibit these signs, chances are that the deliveries are not normal and farmers may have to intervene.

14.4.5 Cold stress and the newborn calf

There can be areas within the tropics, such as in the highlands, where cold stress can adversely affect calf health and performance. Cold stress is called hypothermia.

Wet and cold calves are more prone to hypothermia. Rectal temperature is the most accurate method of determining if a calf is suffering from cold stress. Mild hypothermia occurs when body temperature drops below 37.8°C. Severe hypothermia occurs when it drops below 34.4°C. To combat the cold, calves shiver to increase heat production and they shunt blood from their body extremities to the body core. At 34.4°C, vital organs are cold and impaired brain function results. As core temperatures drop below 30.0°C, signs of life are hard to detect. In extremes, respiration and heart rate drop, animals lose consciousness and die.

Calves can be artificially warmed in various ways: using a warm water bath, warm air or heat lamps and warm blankets. Air movement, such as from a fan, is important to ensure thorough warming of the calf. Calf coats can improve insulation of sick calves, while facilities to heat milk or drinking water can also provide additional methods to warm sick calves.

14.4.6 The 10-point plan for managing the birth process

Davis and Drackley (1998) presented a 10-point plan to manage the birth process as follows:

1. Develop a plan for calving assistance with the veterinarian and all animal caretakers, so everyone knows when to intervene and call in the veterinarian.
2. Provide a sanitary, dry, comfortable calving area with appropriate bedding as necessary. Sanitise the pens after each birth.
3. Monitor all the cows due to calve frequently and regularly and provide assistance promptly if necessary.
4. Ensure the newborn calf's airways are free from mucus and foetal membranes and that breathing is initiated.
5. Dry the calf if the cow is not allowed to do so.
6. Feed ample amounts of good-quality colostrum as soon as possible within the first hour after birth. Do not rely on sucking to achieve the desired intake: use a nipple bottle or stomach tube, if necessary.
7. Clearly identify the calf then separate it from the cow within 12 hr after birth; separate it immediately after birth if concerned about infectious transmissions such as Johne's disease.
8. Dip or coat the navel with 7% tincture of iodine.
9. Maintain strict cleaning and sanitation of all birthing equipment and feeding utensils.
10. Work with the veterinarian to establish and implement an effective vaccination program during gestation and for the newborn calf.

15

The business of calf and heifer rearing

This chapter discusses the importance of documenting all the costs of calf and heifer rearing to be able to prioritise efforts to reduce those that are most expensive.

The main points in this chapter

- Calf rearing is not cheap and producers who try to cut costs will eventually pay for it in the long run.
- Feed costs are generally the most expensive component in calf rearing, but when considering them, it is essential that all existing and alternative feeds are compared on an equivalent basis, such as their supply of dry matter or feed energy.
- An example of costing feeds in the tropics is provided for feeds available in Malaysia.
- A second example is provided for various milk feeding systems in Vietnam.
- The full cost of calf disease is more than just the costs of drugs and veterinary care.
- The full costs of young stock management include variable costs in addition to feed and animal health, and fixed or overhead costs, such as labour, depreciation and sourcing farm finances.
- It is not good farming practice to underfeed replacement heifers. Delayed calving, reduced milk yields and fewer lactations in the milking herd can be the price to pay if calf and heifer growth rates suffer due to poor feeding management.
- Lifetime productivity is very responsive to well-programmed feeding management during heifer growth and milking cow lactation. However, such benefits can be quickly eroded if animal health problems lead to high mortality and stock turnover.
- Farmers feeding lower-quality diets have more to gain, in terms of higher and more secure income, from reducing mortality and involuntary culling than solely from investing in better feeding management.

Calf rearing is not a cheap enterprise, and producers who try and cut costs will eventually pay for it in the long run. Because of the uncertainties of weather in many tropical dairying regions, there can be unusually cold, wet or hot spells and calves are

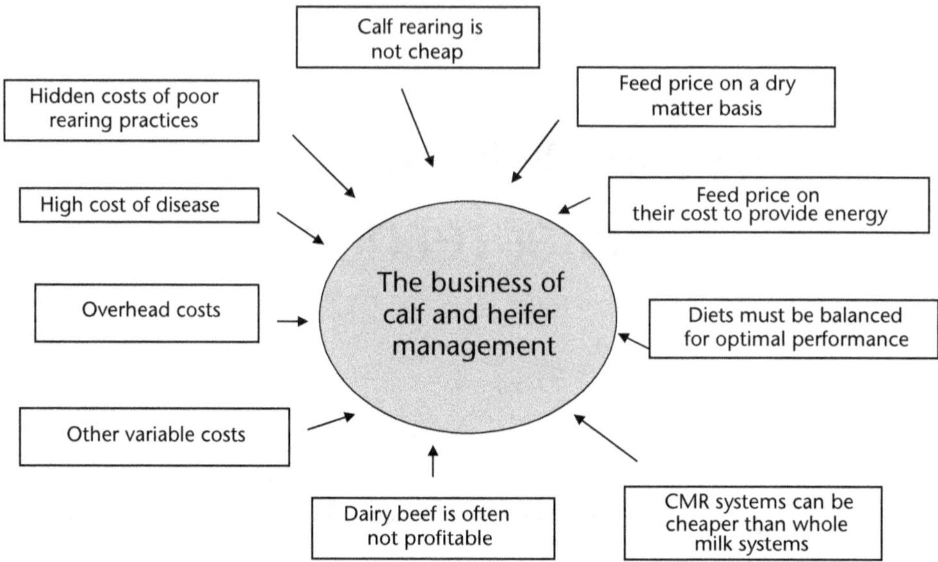

Figure 15.1. The key factors to consider when planning the business of calf and heifer rearing

very susceptible to these changes unless provided with some form of effective and relatively climate-proof housing. Inefficient milk feeding and cleaning systems require more labour and, despite what many producers believe, labour is not free and not even cheap. 'Cut price' milk replacers are generally cheap because they are lower in quality

Newborn calves are an investment to the dairy farm.

than normally priced powders, often because of poor processing techniques. Figure 15.1 highlights the complexity of planning the calf- and heifer-rearing business within the farm's dairy enterprise.

Calves can be reared on less whole milk or milk replacer than is often fed, provided their feeding and management allows for early rumen development. Once calves are weaned, poor feeding practices, such as grazing or being hand fed on low-quality roughages, together with inappropriate concentrate feeding regimes will lead to slow growth. If dairy heifer replacements do not achieve realistic target live weights, long-term milk yields, reproductive performance and longevity will suffer. Sub-optimal growth rates in animals grown for dairy beef increase slaughter ages and can even adversely affect carcass and meat quality. Money spent on good rearing and growing practices will be recouped in improved returns for milk or meat.

Whole milk should not be put into the bulk milk vat for periods of up to 4 or 6 days after calving; recommendations on this vary in different regions. During this period, cows produce colostrum and transition milk, which should all be fed to calves. If only rearing heifer replacements for the dairy herd, the colostrum produced by cows calving down bull and cull heifer calves should provide ample liquid feed for milk-fed calves. Assuming a 25% replacement rate in the dairy herd, together with 45 L of available colostrum and transition milk over several days from each cow, to rear the herd's replacement heifers, this provides 180 L of transition milk for each heifer calf in addition to feeding milk to the remaining calves until their sale at one week of age. This quantity is sufficient to rear a calf from birth to weaning. There should be little need for dairy farmers to buy milk replacers or use marketable whole milk to rear their calves.

This chapter discusses the business of young stock management using recent information from three different tropical countries. Firstly, the feed costs for calf rearing in Malaysia are documented; secondly, the total feed costs for rearing heifer and bull calves in Vietnam are presented; and thirdly, using computer simulation, a detailed evaluation of lifetime productivity of stock on SHD farms in Kenya is discussed in relation to stock mortality and culling procedures.

15.1 Costing different feeds for calf rearing

Various methods for costing whole milk and milk replacers have been described in Chapter 9. Costs can be expressed in terms of local financial units either per kg of dry matter (DM) or per MJ of metabolisable energy (ME) in the product. The latter is calculated from fat and protein levels in the whole milk or milk replacer. Local currency units for various Asian countries are presented in Appendix 3.

Costs for solid feeds, such as concentrates or roughages, can be calculated in a similar manner to liquid feeds once their cost per tonne and their DM and/or ME contents are known. Costs for purchased feeds are easy to calculate, but costs for home-grown feeds are more difficult to determine. Many economists use the opportunity cost of the feed as the basis of its pricing. This is the value of that particular feed if it was sold on the open market. For example, in Australia, wheat can be grown on-farm for say, A$180 per tonne yet could be sold for A$200 per tonne. It should then be priced at A$200 per tonne because that is its actual value to the grower.

The nutritive value of selected feeds has been presented in Chapter 10 in terms of DM, ME and protein, but these can vary considerably for any one feed type. It is strongly recommended when formulating rations for weaned calves, or any livestock for that matter, that actual measures of dry matter, energy and protein be obtained from commercial feed testing laboratories.

Energy- and protein-rich feeds can be purchased already formulated and sometimes pelleted as commercial pellets or they can be blended on farm from the raw ingredients to form a balanced concentrate mix. Calf-rearing pellets often contain vitamin and mineral additives. Commercial pellets, despite being more expensive than on-farm mixtures, are usually the preferred solid feed for calf rearers. Cow pellets are not suitable for rearing calves.

Conserved pasture hay or silage can be priced on its opportunity costs in the open market. However, grazed pasture cannot be priced this way because it has less value as standing feed than when grazed and used by calves. Many farmers undervalue the cost of grazed pasture on their farms. After including the actual cash costs (such as fertiliser, weed control and irrigation), the indirect costs (such as fencing, repairs and maintenance on farm machinery), as well as the costs for labour and depreciation of farm machinery, grazed pasture is not cheap. Some economists even consider council or shire rates and the return on total capital invested in land and equipment when calculating the real cost of grazed pasture. Finally, these are all costs for producing the pasture in the paddock, but it must be remembered that only about half of the pasture grown is actually consumed by grazing stock. At least with cut and carry pastures in the tropics, a greater proportion of that grown (say 70–80%) is converted into a saleable product by the animal.

15.1.1 Feed costs associated with calf rearing in Malaysia

The relative costs of the smorgasbord of feeds available for milk-rearing calves is likely to vary from country to country. Table 15.1 presents a case study for Malaysia in Malaysian ringgits (MR), for various feeds available, using costs in March 2010, and assumed DM, energy and protein values for 'typical' feeds of each type (Brouwer, *pers. comm*). The table includes data on:

- whole milk of two milk fat and protein contents, called good (4.4% fat and 3.6% protein) and average (3.2% fat and 2.7% protein)
- calf milk replacer or CMR (containing 15% fat and 20% protein)
- calf muesli, specially formulated for milk-fed calves
- cow milking concentrate: the usual source of concentrates for milk-fed calves
- palm kernel cake or PKC, the most readily available protein concentrate
- three sources of forage, namely Napier grass, whole crop maize and leucaena leaves.

Because the CMR is less than half the cost of marketable whole milk, it would seem logical to feed whole milk only when it has no market value, such as immediately after calving, and then change to CMR. The calf muesli has been specifically formulated with some long fibre to promote rumen development in the early milk-rearing period. In addition, its high protein content provides for the high protein demands of the young calf. Both the cow milking concentrate and PKC – the usual concentrates fed during milk rearing – contain insufficient protein, hence would delay weaning age.

Table 15.1. Costs for dry matter and energy in various calf feeds in Malaysia (in Malaysian ringgits)

Feed	Dry matter (%)	Energy (MJ/kg DM)	Protein (% DM)	Cost per unit (MR)	Cost for DM (MR/kg)	Cost for energy (MR/MJ)
Whole milk						
Good quality	13.3	23.7	27.1	2.00/L	15.0	0.63
Average quality	11.7	21.8	23.1	2.00/L	17.1	0.78
Calf milk replacer						
Powder	96	18.7	15	6.80/kg	7.1	0.38
Solution	9.1	18.7	15	0.62/L	7.1	0.38
Concentrates						
Calf muesli	89	12	22	3.40/kg	3.8	0.32
Cow milking	85	10	16	1.20/kg	1.4	0.14
Palm kernel cake	85	11	15	0.35/kg	0.4	0.04
Forages						
Napier grass	15	8	11	0.10/kg	0.7	0.08
Whole crop maize	20	8	8	0.10/kg	0.5	0.06
Leucaena leaves	30	8	22	0.30/kg	1.0	0.13

Clearly the lower cost per MJ of energy in calf muesli, compared with whole milk and CMR, would reduce total feed costs through earlier weaning. Even though PKC is the cheapest form of energy of all calf feeds, its lower energy and protein contents and reduced palatability, will limit its potential as a concentrate for young stock. Furthermore, its lack of long fibre will limit its potential to encourage rumen development in milk-fed calves. The calf muesli, even though only recently available, is quickly becoming integrated into successful dairy calf-rearing operations. As demands increase, the cost is likely to decrease, thus improving it economic feasibility for SHD farmers in Malaysia.

Once weaned, the cheaper forages could then become major components of the diet. However, to ensure adequate post-weaning growth rates, the high protein demands of weaned heifers would necessitate incorporating legumes into the forage base and/or feeding high-protein concentrates.

15.1.2 A case study of different milk-rearing systems in Vietnam

Traditionally, in Vietnam, farmers feed fresh milk until weaning at 12 weeks: up to 440 L/calf. In recent times, they have found that recently available CMR is cheaper than whole milk and that calves can be weaned considerably earlier than 12 weeks of age. Rearing male calves for beef is generally not considered economic in Vietnam.

Tiberghien (2009) compared four different calf-rearing systems, each with six male and six female calves. Calves were individually reared in elevated cages until 20 days then in individual pens until weaning at 3 months. They were fed colostrum and transition milk (at 10% of birth weight) for their first 7 days then placed on one of the four feeding systems. They were teat fed the milk or CMR and offered concentrates, hay and water in buckets. The systems were:

Table 15.2. Live weight and growth rates in heifer calves reared on four milk-feeding systems

	Milk 440 L	Milk 10% BWT	CMR standard	CMR EW
Live weight (kg)				
Birth weight	35	31	31	32
3 months	99	92	80	80
6 months	146	143	134	130
9 months	189	186	174	173
12 months	236	230	220	220
15 months	278	282	275	273
Average daily gain (kg/day)				
0–3 months	0.76	0.72	0.58	0.57
3–6 months	0.56	0.60	0.65	0.60
6–9 months	0.50	0.51	0.48	0.50
9–12 months	0.50	0.52	0.55	0.57
12–15 months	0.50	0.61	0.65	0.63
0–15 months	0.49	0.52	0.53	0.52

1. Fresh milk with the amount varying with age totalling 440 L/calf (Milk 400 L).
2. Fresh milk, fixed at 10% birth weight (Milk 10% BWT) then weaned when calves consumed 1 kg/day of starter concentrate.
3. Milk replacer (CMR), standard protocol (CMR standard).
4. CMR and early weaning (CMR EW) then weaned when calves consumed 1 kg/day of starter concentrate.

The research team recorded intakes of milk, CMR and total feed, calf health and live weights up to 12 months (males) or 15 months (females). They also recorded slaughter data for males at 6, 9 and 12 months of age. The growth rate data are presented in Table 15.2. There were no statistical significant differences between feeding systems in live weights at any age and all groups reached the target 270 kg by 15 months of age.

Total feed costs presented in Tables 15.3 and 15.4 were based on the following:

- Whole milk at 7200 VND/L.
- CMR solution at 4500 VND/L.
- Concentrate at 7600 VND/kg.
- Hay at 5700 VND/L.

Table 15.3. Feed intakes and feed costs to 3 months of age for heifers calves reared on four milk-feeding systems

	Milk 440 L	Milk 10% BWT	CMR standard	CMR EW
Milk intake (L/calf)	440	261	54	58
CMR solution intake (L/calf)	–	–	275	347
Concentrate (kg/calf)	58	68	48	51
Hay (kg/calf)	62	70	51	46
Feed costs (000 VND/calf)	**3962**	**2795**	**2281**	**2628**

Table 15.4. Live weight, growth rates and total feed costs in bull calves reared on four milk-feeding systems

	Milk 440 L	Milk 10% BWT	CMR standard	CMR EW
Birth weight (kg)	34	35	35	32
3 months (kg)	92	83	89	79
6 months (kg)	148	135	149	132
9 months (kg)	196	181	197	162
12 months (kg)	**242**	**226**	**247**	**202**
Average daily gain (0–12 months) (kg/day)	**0.57**	**0.53**	**0.58**	**0.47**
Feed costs to 12 months (000 VND/calf)	**8366**	**7407**	**6751**	**7249**

Up to three months of age, the CMR standard system cost only 58% of the Milk 440 L system, but the heifers were 19 kg lighter (80 versus 99 kg). However, by 15 months of age, this weight advantage had disappeared.

The CMR standard system resulted in the fastest growth rates in bulls to 12 months of age followed by the Milk 440 L system. The CMR standard system had the lowest total feed costs to 12 months of age, which were 81% of those in the Milk 440 L system. It is of interest that when expressed as a percentage of total feed costs for bull calves to 12 months of age, the feed costs of heifer calves to 3 months of age ranged from a high of 47% for the Milk 440 L diet to a low of 33% in the CMR standard diet.

One measure of profitability of the bull calves in the various systems, namely the carcass return less total feed costs, are presented for all systems in Table 15.5 and for the CMR standard system in Table 15.6. The carcass return is based on 90 000 VND/kg meat. The only profitable rearing system for slaughter calves was CMR standard. Profit decreased with advancing slaughter age mainly because the increase in feed costs was greater than the increase in carcass value.

The CMR standard rearing system has also been assessed on a Vietnamese commercial dairy farm where the calves grew at 0.48 kg/day over their first 3 months with total feed costs of 2.134 million VND, compared with 0.55 kg/day and 2.329 M VND respectively on the above trial.

Tiberghien (2009) concluded that, although the calves grew more slowly on the CMR standard system, total feed costs were lower than when fed whole milk, and by 15 months of age, any weight advantage disappeared. In addition, with the low meat returns in December 2009 for slaughtered calves in Vietnam, the CMR standard system was the only one profitable for rearing dairy beef.

Table 15.5. Profitability of rearing bull calves for slaughter at three ages on four milk-feeding systems (000 VND/calf)

Slaughter age (months)	Milk 440 L	Milk 10% BWT	CMR standard	CMR EW
6	–1356	–678	1165	–688
9	–1285	–1226	835	–995
12	–981	–539	768	–1117

Table 15.6. Profitability of rearing bull calves for slaughter at three ages on the CMR standard system

	Slaughter age (months)		
	6	9	12
Slaughter live weight (kg)	149.1	197.2	247.3
Meat (%)	37.5	34.1	33.8
Carcass value (000 VND/calf)	5031	6039	7524
Total feed costs (000 VND/calf)	3866	5204	6756
Profit (000 VND/calf)	1165	835	768

15.2 Other costs for calf and heifer rearing

Surveys conducted in the US and the UK found that feed accounted for only 50–60% of the total costs for raising heifer replacements to first calving. Additional costs that should be considered include:

- other variable costs such as veterinary and drugs, replacing feeding equipment, replacing bedding, mating (using bulls or semen), fuel and electricity, and death losses
- fixed or overhead costs such as labour, depreciation of facilities and interest on finances specifically borrowed for the calf- and heifer-rearing enterprise.

15.2.1 The high cost of diseases in calves

The losses through disease during calf rearing can be classified as follows:

- deaths, hence loss of calf value with little (or usually no) salvage value for the carcass
- costs of veterinary services plus drugs
- costs of extra feed required when calves lose or do not gain weight when sick, hence require more feed to reach target live weights
- costs of transport and resale of any calves culled
- costs of reduced throughput in rearing unit, additional labour for treatment and greater interest on loans, etc.

Veterinarians in the US have calculated that each sick calf requires on average, 53 min of extra care before recovery occurs. In terms of labour, veterinary services and drugs, the cost for each sick calf was at least US$18. Good calf rearing and husbandry and sound economics must then go hand in hand.

15.3 The hidden costs of poor heifer rearing

It is not good farming practice to underfeed replacement heifers. Growth rates should be maintained between weaning and first calving, otherwise heifers will not reach their target live weights for mating and first calving. Undersized heifers have more calving difficulties, produce less milk and have greater difficulty getting back into calf during their first lactation. When lactating, they compete poorly with older cows for feed and because they are still growing, will use feed for growth rather than for producing milk. They are more likely to be culled for poor milk yield and/or infertility.

Well-grown heifers require a lot of feed. For example a 12-month-old heifer can consume 55–60% of the feed consumed by a milking cow producing 20 kg/day of milk. Adopting low-input feeding programs could decrease heifer rearing costs. These include reducing the amount of concentrates and hay fed or only feeding heifers on poor-quality forages. Furthermore, 60% of the cost of heifer rearing occurs during the first year when management demands are higher.

As previously discussed, low live weights at mating can lead to poorer conception rates, increased calving difficulties and reduced fertility in later years. Life-time performance of poorly managed heifers can be adversely affected in at least three ways:

1. The potential milk production of dairy heifers is reduced by any increase in age at first calving. For example, if calving age was increased from 24 to 30 months and heifers produced 4000 L in their first lactation, each heifer would forgo 2000 L of milk.
2. Assuming a herd replacement rate of 25% for a 50-cow herd, delaying calving by 6 months would result in an extra seven non-productive heifers competing with the milking herd for resources. If these resources were used to feed more cows, an extra three cows could be milked. Assuming cows produced 4000 L per lactation, this amounts to a loss of 6000 L (4000 L × 0.5 lactation × 3 cows) or 857 L per heifer.
3. Failure to reach target live weights at first calving can reduce lifetime milk production by up to 20 L/kg below target live weight at first calving. This amounts to a potential loss of, say, 1050 L per heifer with 50 kg reduced live weights.

Therefore, if rearing costs were reduced to such an extent that heifers calved at 30 months weighing 500 kg, rather than 24 months weighing 550 kg, this can decrease their potential milk yield by 3950 L per heifer.

There are other less obvious costs in later calving, such as slower rates of genetic progress, older herd age structure, hence potential to increase the incidence of age-related problems (such as mastitis and infertility), and increasing involuntary culling, hence reduced opportunities for selection on milk production.

Therefore, when rearing replacement heifers, farmers must plan heifer feeding carefully because, in the long run, underfeeding them can cost a lot of money.

15.4 Measures of lifetime productivity on small holder dairy farms

Undertaking long-term studies of the key on-farm factors influencing lifetime productivity of dairy cows on small holder farms would require numbers of researchers and resources rarely available, if at all, in the tropics. Through the use of computer simulation, Rufino *et al.* (2009) have developed a dynamic model to undertake such a detailed evaluation of the impact of various feeding and herd strategies on the performance of crossbred dairy cows under traditional and improved production systems on small holder farms in Kenya. They simulated four different feeding systems, as follows:

- System 1: *Ad lib* Napier grass with no supplements
- System 2: *Ad lib* Napier grass with strategic supplements of maize silage for 6 months of the year

- System 3: *Ad lib* Napier grass plus strategic maize silage supplements together with 2 kg/day of formulated concentrates throughout the entire lactation
- System 4: *Ad lib* Napier grass plus strategic maize silage supplements together with optimal concentrates, namely 0.5 kg/day of formulated concentrates during calf and heifer rearing and during the dry period, 5 kg/day during the first 150 days of lactation and 1 kg during the rest of the lactation.

Being a series of computer simulations, the authors had to make various assumptions about farm management and cow performance, all of which were based on locally derived research findings. These included:

- Cows lived until 13 yr of age with mature live weights of 500 kg.
- Calves were weaned at 4 months of age.
- Cows had maximum milk yields of 14.6 kg/day and 4450 kg/lactation.
- Cows had 10 month lactations and were dry for 2 months.
- Calving rates varied from 25% to 90% depending on cow live weight loss and body condition.
- Mortality rates for calves up to 3 months of age were 15%.
- Mortality rates for cows were 7% (between 2 and 6 yr) and 12% (between 7 and 13 yr).

Rufino *et al.* (2009) repeated the computer analyses using two different scenarios, firstly without, and secondly with, the various assumptions for mortality rates (which

Table 15.7. Influence of feeding management on indicators of lifetime productivity in two different scenarios, namely with (+) or without (–) assumed mortality and culling rates, on small holder dairy farms in Kenya (see text for details of each feeding system)

Indicators	Mortality and culling rates	System 1	System 2	System 3	System 4
Lifetime (yr)	–	13.0	13.0	13.0	13.0
	+	7.8	7.3	8.1	7.9
Productive life span (yr)	–	9.4	8.9	9.0	9.9
	+	4.4	4.0	4.0	5.0
Calves (no./lifetime)	-	5.8	5.2	5.6	7.3
	+	3.0	2.8	2.9	3.9
Cumulative milk yield (t/lifetime)	–	14.6	10.7	17.0	25.4
	+	7.5	3.7	8.2	14.4
Days in milk (% per lifetime)	–	35	33	35	45
	+	28	25	26	37
Daily milk yield (kg/day)	–	8.5	7.2	10.6	12.0
	+	8.3	5.3	9.9	11.9
Cumulative net income* (% relative to System 1)	–	100	80	122	180
	+	54	27	59	102
Milk yield (kg/lactation)	–	2500	1900	3000	3500
Days open (days/parity)	–	382	363	358	254
Age at first calving (months)	–	43	48	48	37
Calving interval (months)	–	20	20	19	14

* cumulative net income takes into account milk returns, concentrate purchases and sale values of bull and heifer calves.

also included involuntary cullings) for stock of different ages. The indicators of lifetime productivity, using these two scenarios are presented in Table 15.7.

15.4.1 Scenario 1. Assuming zero mortality and culling

Providing maize silage with targeted concentrate feeding had a dramatic impact on both milk yields (per lactation and over the entire productive life) and cumulative net incomes. Feeding concentrates during calf and heifer rearing markedly reduced age at first calving and days open, which can be quantified as days in milk over the entire lifetime, which was up to 26% higher in System 4. Even though the total productive life span was only up to a year longer, the cows in System 4 calved up to 7 months earlier and had calving intervals up to 6 months shorter. Net farm returns were up to 80% higher in System 4, clearly indicating that improved feeding management is highly profitable.

15.4.2 Scenario 2. After taking into account the impacts of mortality and culling

Only 28–31% of stock survived for 13 yr. Productive life spans, calves produced and cumulative milk yields are all dramatically reduced once stock wastage (mortalities and involuntary cullings) were taken into account. These productivity indicators were reduced by between 43 and 65%, depending on the feeding system. The impact of animal health was such that the cumulative net income of the most profitable System (4) was reduced to that of System 1 with no stock losses. In other words, the effect of improving feeding management was completely negated by the 49% reduction in length of productive life span. Associated studies in Tanzania (Ngategize 1989) concluded that the benefits of increasing survival by 5% (namely higher milk production, higher off take and higher capital value) exceeded the costs in implementing a disease-control program.

15.4.3 Conclusions

The fact is that farmers feeding lower-quality diets have more to gain, in terms of higher and more secure income, from reducing mortality and involuntary culling rates than solely from investing in improved feeding management. This emphasises the importance of tackling probably the major constraint of animal health: namely during the calf- and heifer-rearing phase.

Calving at an early age and short calving intervals should be major goals in small holder dairy farming, otherwise farmers cannot achieve the returns of their large investments in animal and farm capital. If optimised diets are used without reducing current mortality rates, farmers are prevented from earning higher and more stable incomes. Therefore, improving lifetime productivity requires investments in both improved feeding management and reduced disease-related mortality rates.

16

Assessing current calf- and heifer-rearing practices

This chapter discusses a process to assess current practices and grade farmer skills in young stock management.

The main points in this chapter

- The first step in improving the on-farm management of calf and heifer rearing is to audit the current system (only then will it be possible to develop a modified program to overcome any observed deficiencies).
- The audit can be separated into:
  - Peri-natal (pre-calving and first 24 hr)
  - Pre-weaning feeding management
  - Pre-weaning herd and health management
  - Weaning process
  - Weaning to first calving.
- The farmer's skills can also be graded using both objective and subjective criteria.
- A short list of high-priority management skills distils this assessment down to the 10 most important practices in young stock management.

The first step in improving the on-farm management of calf and heifer rearing is to audit the current system (see Figure 16.1). Only then will it be possible to develop a modified program to overcome any observed deficiencies. The following provides a series of checklists to allow current management practices to be assessed. In addition to the complete list containing 50 series of observations, a 'high priority list' of just 10 observations has been developed.

16.1 Checklists for assessing the system

16.1.1 Peri-natal (pre-calving and first 24 hr)

- pre-calving facilities and hygiene
- separation of calf from dam

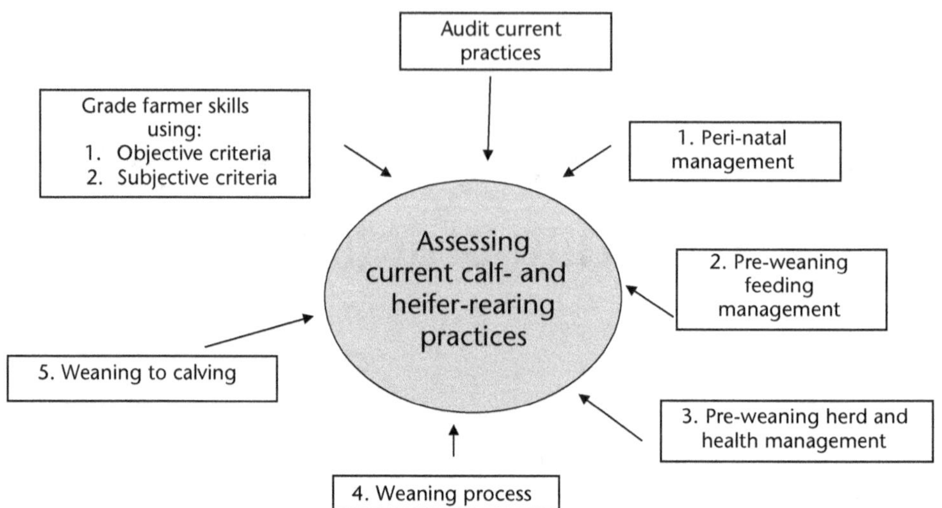

Figure 16.1. The key processes to consider when assessing current young stock management

- age of calf when drinking first colostrum
- colostrum feeding (awareness of how best to obtain benefits from colostrum, putting theory into practice)

Workshop participants assessing a farmer's facilities for milk rearing his dairy heifer replacements.

- calf identification
- housing of pre-weaned calves (stalls, bedding, separation from adult stock, cleanliness).

16.1.2 Pre-weaning feeding management

- development of feeding programs to promote rumen development
- continual access to clean water
- type and method of liquid feed (feeding frequency, method of feeding, daily feeding rate)
- type of concentrate (age when commence feeding, quantity fed, formulation)
- type of forage (age when commence feeding, fresh versus hay, forage quality).

16.1.3 Pre-weaning herd and health management

- development of animal health program
- pre-weaning calf mortality
- condition of the milk-fed calves (body condition, signs of ill-thrift)
- prevention and treatment of scours (electrolytes, identification of causative agent, effective use of antibiotics)
- prevention and treatment of pneumonia
- prevention and treatment of other animal health issues (such as lameness or climatic stress)
- skills with identifying and nursing sick calves
- routine husbandry practices (debudding, vaccinations, drenching, ectoparasite treatment, castration)
- drug storage
- hospital pens
- veterinarian support (when sought, quality of support)
- document pre-weaning morbidities (sickness) and mortalities (deaths)
- fate of bull calves
- sources of labour (owner, family, employed and their efficient usage)
- shed environment (effluent management, ventilation, coping with heat stress)
- general stock welfare and handling facilities/skills
- record keeping (system, feed inputs, animal health, labour)
- calculate total pre-weaning costs
- general hygiene (cleaning of feeding equipment, calf pen and shed hygiene).

16.1.4 Weaning process

- basis of weaning (age, weight, concentrate intake)
- movement to weaner pens.

16.1.5 Weaning to first calving

- continual access to drinking water
- shed environment (effluent management, ventilation, coping with heat and cold stress)

- feeding management (forages and concentrates)
- chopping forages
- feed storage facilities
- monitoring feed intakes, weight, body condition, heat stress (stock and shed)
- post-weaning heifer mortality
- condition of stock (body condition, signs of ill-thrift)
- age of first oestrus
- criteria for mating (age, weight)
- mating program (bull versus AI)
- routine pregnancy testing
- preparation for calving (pre-calving facilities and hygiene)
- calculating total rearing costs
- fate of weaned bulls
- age at first calving
- wastage rate from birth to second calving
- subjective assessment of overall farm management skills.

16.2 Grading farmer skills in young stock management

A structured approach can be used as a framework to assess young stock management more objectively and also to provide a focus for group discussions. For example, the

Advisers from Indonesia interviewing a farmer about his herd management.

following set of criteria can be developed using a grade of good, average or poor, with some objective and others subjective.

16.2.1 Objective criteria for grading skills

- pre-weaning morbidity (requiring drugs and/or veterinary assistance (%)
- pre-weaning calf mortality (%)
- daily milk feeding rate (L/calf/day)
- weaning age (days)
- age at first mating (weeks/months)
- age at first calving (weeks/months)
- wastage rate (from birth to second calving) (%)
- record keeping
  - if the farmer writes the details down and remembers, good
  - if he remembers but does not write them down, average
  - if he does not know or remember, poor.
- productivity of first-calf heifers (in L milk over first lactation).

16.2.2 Subjective criteria for grading skills

- condition of milk-fed and weaned stock
- condition of calf-rearing pens and facilities.

16.3 A shortlist of high-priority calf and heifer practices

If an adviser visited a small holder farmer with the above series of 50 observations to be made on his farm, the farmer could find such an exercise intimidating and too intrusive and the adviser may have difficulty completing the task. The following shortlist of high-priority management skills distils this down to the 10 most important practices in young stock management.

1. colostrum feeding management
2. type and method of liquid feed
3. continual access to clean water
4. type of concentrate
5. basis of weaning
6. development of animal health program
7. pre-weaning calf mortality
8. shed environment
9. post-weaning heifer mortality
10. age at first calving.

17

Conducting training programs on improved young stock management

This chapter presents a framework for workshops on improved young stock management on small holder farms.

The main points in this chapter

- The target audiences for these workshops can be advisers or farmers, or preferably a combination of the two.
- It is important to develop a clear set of workshop objectives so all participants know why there are there. The key learning outcomes can be selected from this manual.
- Asking participants to complete an expectation form at the beginning and an evaluation form at the end of each workshop, in their local language, helps plan each day's program and provides valuable feedback.
- Distributing copies of all overheads, translated into their local language, prior to the workshop will improve comprehension of the material as it is being presented.
- With farmers, the emphasis should be on visual images followed by practical examples of poor versus good farm practices.
- Farms to visit should be chosen with a specific purpose, such as to demonstrate a particular farming practice or set of practices.
- Conducting group presentations, where small groups prepare and report back on specific aspects of the program, encourages active workshop participation.
- CalfTrack is a comprehensive training program for calf-rearing staff developed by Pennsylvania State University.
- CalfTrack contains a comprehensive health action plan.
- There are many other technical resources freely available on the internet.

This manual provides the topics that could be used in a workshop program specifically for farmers planning to improve their young stock management practices, using Figure 17.1 as a framework. However, farmers, being practical people, learn more from seeing and

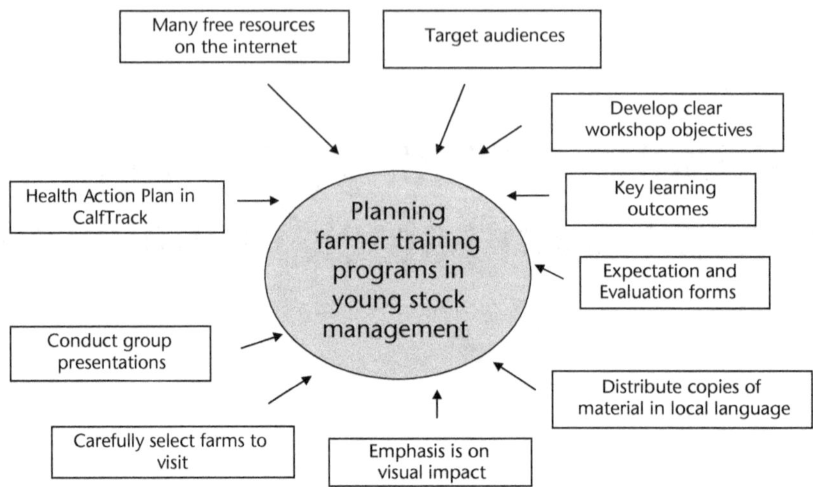

Figure 17.1. The key elements of workshops on improved young stock management

doing rather than listening and reading. This is called experiential learning. Therefore, to improve their understanding of the principles discussed in this manual, farmers need to be provided with practical examples of these improved management practices.

One key feature of the dairy industry in Australia is regular farmer meetings, where groups of farmers visit other farms, particularly those that have proven to be successful in adopting improved farm practices. To be successful, such 'farmer discussion groups' need a clear focus and set of objectives, such as:

Assessing the age of a dairy heifer in Indonesia from its teeth development.

- Why are the discussion groups necessary?
- How is there empowerment of group members and cooperative development within the group?
- How is the subject matter chosen?
- What are the most effective approaches to technology transfer?
- How can each member be guaranteed that they can equally provide input and gain benefits from the meetings?

Such topics are generally outside the scope of this book, but need to be considered to ensure each farmer is rewarded for his time and effort at the group meeting.

The first phase in developing a focus could be to conduct a workshop on improved young stock management on SHD farms. This chapter discusses the planning and implementation of such a workshop to encourage adoption of farm practices to exploit the productive potential of dairy replacement heifers more fully. Such a workshop should include both local advisers and dairy farmers, so advisers can more easily develop follow-up meetings and maybe facilitate a series of farmer discussion groups.

17.1 Planning farmer workshops

17.1.1 Workshop objectives

1. To provide an understanding of the effect of calf and heifer rearing on long-term dairy cow performance.
2. To understand more fully the key principles behind the transfer of passive immunity via colostrum and rumen development in milk feeding systems for replacement dairy heifer calves.
3. To understand more fully the importance of cleanliness and hygiene during milk rearing on the health and welfare of calves.
4. To understand more fully the importance of setting and achieving growth targets during the post-weaning period of heifer rearing.
5. To develop a simple set of relevant checklists to ensure greater success in young stock management, hence improved profits from the milking herd.

17.1.2 Developing a workshop program

Prior to visiting local farms, the participants should be exposed to some of the theories behind the reasons for failure of previous attempts to improve the productivity of replacement heifers on small holder farms. Table 17.1 summarises the topics, discussed in previous chapters of this manual, some or all of which could be considered in the classroom sessions.

In addition to some or all of the pre-determined topics listed in Table 17.1, a session should be incorporated into the last day of the workshop to discuss those topics listed in the expectation forms from workshop participants. It is important that these be covered because farmers (and advisers) may have particular practical issues that they hope can be solved during the workshop.

Table 17.1. Suggested topics for a workshop on improved young stock management

Introduction	
Importance of well-grown heifers to Asia's tropical dairy industry Cost of poorly reared heifers The five key objectives of rearing heifer replacements Measures of success in calf and heifer rearing	
Rearing calves to weaning	***Growing out weaned calves to first calving***
Newborn calf management Close up cow care Cleanliness of maternity pen Observing newborn calves Calving problems and assistance Navel cord dipping Identification and ear tagging Weighing and measuring	***The benefits of well-grown heifers*** Improved milk production Improved fertility Reduced age at first calving Reduced dystocia Reduced wastage
Colostrum feeding management Quality testing Antibody transfer Colostrum storage Feeding colostrum Oesophageal feeder	***Growth targets*** Target milk yields Target live weights Body condition Target wither heights Calculating relative milk yields
Milk feeding management Checking milk replacer quality Mixing milk replacer Feeding whole milk Feeding method	***Effect of age at first calving on farm profitability*** Reduced rearing costs Improved reproductive performance Greater longevity, hence lifetime production Reduced need for replacements More rapid increase in herd size Effects on farm profit
Cleaning and sanitation Cleaning equipment Sanitising equipment Cleaning and disinfecting housing Chemical handling and safety	***Mating management*** Onset of puberty Natural mating Bull management Artificial insemination Heat detection
Dry feed and weaning Quality calf starter Feeding calf starter Feeding water Weaning healthy calves	***Feeding heifer replacements*** Nutrient requirements of growing heifers Testing feed quality Providing adequate forages Supplementing with specific nutrients Critical growth periods Off-farm rearing schemes

Calf comfort and housing Housing evaluation Bedding maintenance Handling and restraint	***Disease prevention and health management*** Preventing diseases through vaccination Internal parasites External parasites Mastitis Johne's disease Nutritional diseases Developing a health management protocol
Communicate with your calves What makes a good calf rearer? Using all your senses Understanding calf 'language' Visual signs of calf stress and sickness How calves react to people Calf welfare	***Economics of heifer rearing*** Costs of farm inputs, including labour Economic implications of reducing farm inputs Dairy beef production
Calf health Environment assessment Daily health checks Identifying calf sours Identifying respiratory diseases Evaluating general appearance Taking a calf's temperature Dehydration and electrolyte fluid replacers Giving injections and handling vaccines Dehorning and removing extra teats What do you do with sick calves? Maintaining a healthy calf shed	***Developing a checklist for rearing heifers*** The 'golden rules' of heifer rearing Recording the most important results of good management Best management practices in heifer rearing
Economics of calf rearing Cost of feed inputs Other costs of calf rearing The high cost of disease What makes a good calf-rearing system?	
Developing a checklist for rearing calves The 'golden rules' of calf rearing Recording the most important results of good management Best management practices in calf rearing	

In a 3-day workshop, the farm visits could be planned for day 2, to break up the program and provide a full day to discuss the key aspects of improved young stock management. Associated with the farm visits could be a series of small group presentations in which several participants could be given a specific topic to discuss with the farmer being visited, asked to spend time on returning to the classroom to prepare a short (5- or 10-min) presentation, then deliver it in front of all the participants. This will focus the participants on a specific farm activity during the farm visit. For example, different groups could be asked to assess the farmer on the five key topics of a farm audit on young stock management as presented in Chapter 16 of this manual.

17.1.3 Practical issues when planning the workshop program

The workshop material will differ depending on the key audience and following discussions with extension specialists in the various countries for which the program is to be designed. As well as a simple set of PowerPoint presentations, technical workshops for government advisers and trained herd managers could include distribution and discussion of this manual.

Examples of expectation forms to be filled in by each participant at the beginning, and evaluation forms at the completion of workshops are presented in Appendix 4. These are very important, firstly, because workshop participants may not all be aware of the workshop's emphasis on improved young stock management practices and, secondly, this can help plan an 'open session' on the last day to discuss specific issues on dairy production technology.

Prior to the workshop, each participant should receive copies of all the PowerPoint overheads translated into their local language. This is important for ease of comprehension during their presentation and to provide pages for writing down additional notes during the workshop. At the close of each workshop, participants could each be presented with a certificate of attendance.

During the workshops, participants can be offered small gifts, such as Australian souvenirs, as rewards for individual oral presentations or as gifts for farmers who opened up their farm (and books) for participants to visit. Government workshop organisers could also receive similar gifts.

It is important to involve participants in the workshops in addition to them listening to and discussing course material. Conducting group presentations, where small groups prepare and report back on specific aspects of young stock management, encourages active participation in the workshops. It also provides opportunities for public speaking, which many may have not been asked to do previously. In addition, it gives a 'local flavour' to the workshop, which is very valuable for the presenters as well as to the participants.

17.1.4 Associated farm visits

A key element of any farmer workshop is a series of farm visits, preferably to farms with different levels of young stock management, so participants can see for themselves what constitutes poor and good farm practices. It is one thing to develop a series of best management practices (BMP), and another thing to see attempts to adopt them, but, of greatest importance, understand the reasons why farmers fail to achieve them. Failure could be due to:

- ignorance: farmers don't know about them
- lack of resources: farmers don't have the money or facilities to adopt them

Dairy advisers from Thailand collating information from a farm visit.

- lack of incentives: farmers don't see the need to want to adopt them
- lack of service provided by government advisers or local agribusiness providers.

Unfortunately, many of these BMPs are like insurance policies in that farmers take the attitude that 'if it isn't broken, you don't need to fix it', particularly if it costs money to modify facilities or purchase additional equipment. Such issues should be discussed back in the classroom following the farm visits.

It is important to select farms to visit based on the high quality of their young stock management and the good record keeping of the farmer. Farmers will learn much from seeing a well-managed calf shed and heifer unit, particularly when the farmer can discuss the costs and returns from his system.

Including a veterinarian in the group will allow more discussion of the major diseases affecting young stock, with the aim of developing a simple set of guidelines for BMP, much the same as the 'golden rules' discussed at the workshop.

Selection of suitable farms is therefore paramount when planning the visits. Close proximity to the workshop venue is important. Each farm should be chosen with a specific purpose, such as to demonstrate a particular farming practice or set of practices. Ideally, farmers should have good records so they can provide useful background on these observed practices, such as their costs of calf and heifer rearing. Selected farmers should obviously not be intimidated by large groups of inquisitive visitors. Inviting farmers back to the workshop is useful so they can further explain their management decisions during the debriefing session.

As previously mentioned, forming small groups of workshop participants with specific tasks allows a lot more information to be collected on farm than if the visit was less structured. These groups could meet back in the classroom to prepare a short presentation summarising their observations and interpretations.

Throughout South and East Asia, management of young stock rarely receives adequate attention, with the acceptance of very high calf and heifer mortality (15–25%) as 'normal'. Therefore this is one area where a little more attention to detail can pay large dividends. Such a change in approach should naturally follow on from a better understanding of the theories behind, and the practicalities of, improved calf and heifer rearing.

17.2 CalfTrack: calf management training system

CalfTrack is a comprehensive training program developed by Dr Jud Heinrichs and his team from Pennsylvania State University (Heinrichs 2002). The training package provides a series of documents, namely:

- **Trainer's guide:** a 70-page comprehensive review of calf-rearing principles.
- **Chore plan:** a loose-leaf set of instructions for the many tasks of calf care.
- **Score guide:** this provides score cards for different descriptions of scours, respiratory observations and general calf appearance (see Table 17.2) to develop a colour-coded 'total daily score' for the health status of each calf. It also allows for the weekly recording of the average daily concentrate intake.
- **Individual calf records:** this allows for recording the animal health observations for each calf every day for 8 weeks, together with space for managers', veterinarians' and employees' hand-written notes.
- **Compact disc of all the printed material:** this allows for the printing of blank individual calf records for each calf.
- **Orientation video:** this outlines the entire training program, as well as providing a 25-min video with excellent visual descriptions of many aspects of calf care.

The training program is separated into seven components, namely:

1. Newborn calf management.
2. Colostrum management.
3. Liquid feed management.
4. Cleaning and sanitation.
5. Dry feed and weaning.
6. Calf comfort.
7. Calf health.

Key features of some of these components are presented below.

17.2.1 Newborn calf records

The following information can be recorded for each calf after it is born:

Calf information

- initials of staff recording the data
- calf's and dam's identification numbers
- date and time of birth
- calving difficulty from 1 to 5; 1, no assistance; 2, minor assistance; 3, hard manual assistance; 4, mechanical extraction; 5, caesarean section
- birth weight
- if the navel was dipped in iodine solution
- date and live weight when calf was weaned.

Feeding information

- colostrum feeding: time fed, quantity, quality, fresh or frozen
- amount of milk or milk replacer to feed
- amount of starter grain to feed.

17.2.2 Animal health records

The coded scoring system allows staff to assess the health status of any calf quickly and accurately using the descriptors in Table 17.2.

The total daily score is then calculated from the sum of the three scores in Table 17.2. In addition, each calf can be allocated a daily colour code as shown in Table 17.3.

The colour code provides a daily visual indication of the health status of each calf, which can be used to plan further veterinary action as described in Figure 17.2. Based on this action plan, the 'follow ups' can range from:

Table 17.2. Descriptors for scoring calves on scours, respiration observations and general appearance

Score	Scours	Respiratory observations	General appearance
1	Normal consistency, brown to light colour, normal odour	Normal (no cough, slow breathing)	Normal, alert, bright eyes, ears up
2	Soft to loose consistency, yellow, brown or green colour, mucus, slight odour	Slight cough, runny nose, watery eyes, slow and normal breathing	Slightly off, droopy ears
3	Loose to watery consistency, yellow or green colour, mucus, strong odour	Moderate cough, runny nose, watery eyes, rapid breathing	Moderately depressed, head and ears droop, dull or sunken eyes, lethargic
4	Watery consistency, yellow, green or clear colour, mucus, slight blood, strong odour	Moderately severe and very frequent cough, mucus discharge from nose, watery eyes, rapid panting	Moderately severe depression, head and ears droop, dull sunken eyes, will not rise
5	Watery consistency, clear colour, mucus, bloody	Severe and chronic cough, irregular, weak to rapid breathing, eyes rolling, mucus discharge from nose	Severe depression, flat on side

Table 17.3. Allocating a colour code to the daily scores

Sum of three scores	Total score	Colour code
4 or less	1	Green
5	2	Blue
6	3	Yellow
7	4	Orange
8 or more	5	Red

- no further action required
- farm staff providing heat lamp or blanket for calves with body temperatures below 38.1°C
- farm staff administering fever reducers for calves with body temperatures above 39.2°C
- farm staff administering electrolytes to treat dehydration
- veterinarian administering antibiotics for severe scours or respiratory problems.

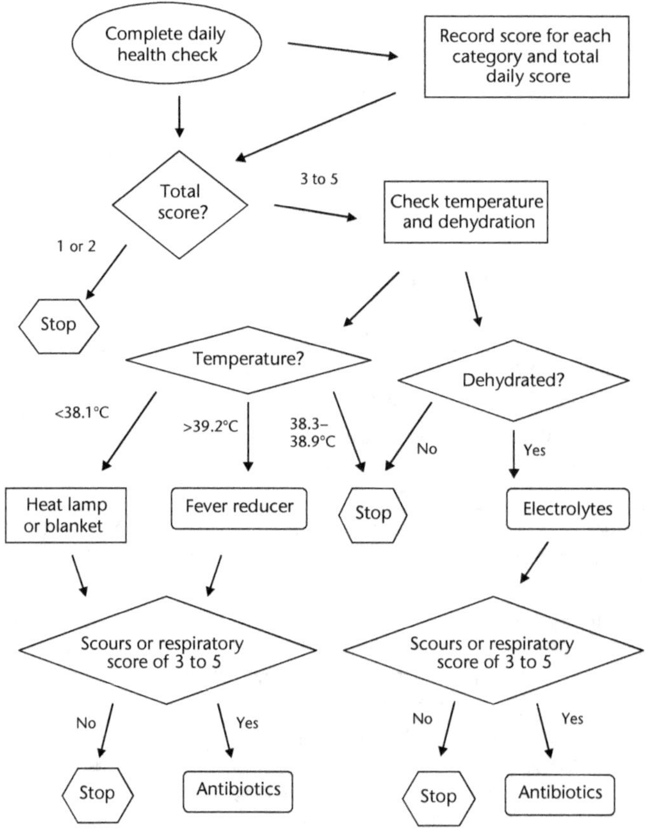

Figure 17.2. The CalfTrack health action plan based on individual scores for scouring, respiratory observations and general calf appearance

17.3 Additional training material on the internet

There is considerable technical material on calf and heifer rearing available on the internet. Much of it is free and can be regularly subscribed to.

17.3.1 Attica Veterinary Association

This is an American veterinary website which makes freely available, excellent monthly newsletters dealing specifically with dairy calf feeding and management. In addition to back issues of the newsletters, the website contains a link to sourcing a CD on calf rearing, called 'Calf Manager' at <http://www.atticacows.com/orgMain.asp?orgid=30&storyTypeID=&sid=&>.

Calf Manager contains many newsletters and booklet on calf rearing, which include:

1. Calving Ease (56 articles written by Dr Sam Leadley): <http://www.atticacows.com/orgMain.asp?orgid=11&storyTypeID=&sid=&>.
2. Calf Management Facts (87 articles written by Dr Sam Leadley): <http://www.atticacows.com/orgMain.asp?orgid=19&storyTypeID=&sid=&>.
3. Calf Notes (108 articles written by Dr Jim Quigley): <http://www.calfnotes.com>.
4. Feeding the newborn dairy calf, an extension manual written by staff at Pennsylvania State University (Heinrichs and Jones 2002): <http://www.cas.psu.edu>.

17.3.2 US Dairy Calf and Heifer Association

This association has a weekly newsletter as well as many technical bulletins: <https://calfandheifer.site-ym.com/>.

17.3.3 Calf and Heifer Adviser

This newsletter is produced by the US magazine *Dairy Herd Management*. It can be sourced at <www.dairyherd.com/adviser/subscribe.htm>.

It contains an extensive library of calf and heifer technical material.

17.3.4 Dairy Australia

Dairy Australia has an extensive collection of calf-rearing material. It is all freely available at <http://www.dairyaustralia.com.au/Animals-feed-and-environment/Animal-welfare/Calf-welfare/Rearing-healthy-calves-manual.aspx>.

18

Best management practices for rearing young stock

This chapter introduces the concept of best management practice to audit farm practices in rearing dairy heifer replacements.

The main points in this chapter

- A US study of calf rearing recorded ten calf-rearing practices that were closely associated with mortality in milk-fed calves.
- Best management practice and quality assurance are processes for describing and implementing the most suitable procedures for a particular set of tasks to achieve a desirable outcome.
- Essentially it means 'Saying what you should do', 'Doing what you say' and 'Recording what you have done'.
- With regards rearing dairy heifer replacements, it can be undertaken by auditing the six major components of any heifer-rearing program. Checklists have been developed for these six, namely:

1. Planning general herd and heifer management.
2. Planning heifer supply programs.
3. Planning heifer care from birth to weaning.
4. Planning heifer care from weaning to mating.
5. Planning heifer mating programs.
6. Planning heifer care from mating to calving.

'Best management practice' and 'quality assurance' are processes for describing and implementing the most suitable procedures for a particular set of tasks to achieve a desirable outcome. With something as complex as running a dairy farm, it is best to partition the major outcome – that is, profitable milk production – into several sets of management decisions that producers must make. These include growing productive forages, efficient feeding management, effective animal health and rearing replacement heifers (see Figure 18.1). Essentially best management practice means:

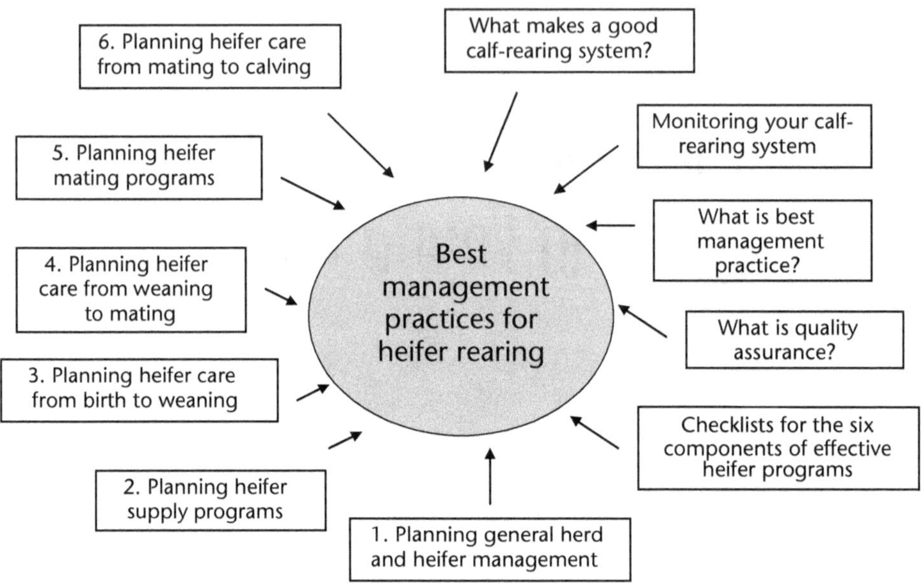

Figure 18.1. Best management practices for rearing dairy replacement heifers

- saying what you should do
- doing what you say
- recording what you have done.

This should improve all aspects of your calf-rearing operation.

Quality assurance (QA) programs are in their infancy in Asia. With increasing focus on meeting customer requirements, QA programs are likely to become an integral part of dairy enterprises in future years.

18.1 What makes a good calf-rearing system?

For every hundred dairy farmers, there would be close to a hundred different ways they rear their heifer replacement calves. In Western countries, more and more farmers are feeding less milk for fewer weeks and reaping the benefits of early rumen development, provided they feed top-quality concentrates and low-quality, but palatable, roughages. Most farmers now provide shedding for their milk-fed calves to protect them from the extremes of climate. There may be only 20 different rearing 'systems', but a lot more rearing practices, after taking into account the subtleties of physical facilities, feeding programs, disease management and human–calf interactions.

A survey of 1700 dairy farms in the US documented a series of calf-rearing practices that were closely associated with mortality in milk-fed calves (Losinger and Heinrichs 1997). For a variety of reasons, US dairy farmers lose many more pre-weaned calves (8–9%) than do Australian farmers (3–5%), whereas, from Chapter 6, calf mortalities on tropical Asian dairy farms are likely to be 15–25%. The following list summarises those factors that were associated with high death rates in US calf-rearing units in year-round calving herds. It must be emphasised that these data are for 'typical' US dairy systems that differ in many ways to Asian farms. Therefore, it is the general principles rather than the actual 'numbers' that are important in this study.

18.1.1 Herd performance

Low-producing herds have higher calf losses. In the study, low production was quantified as less than 7700 L/cow rolling herd average, which would equate to herds producing less than 3000 L/cow rolling herd average in Asia. One concludes that lower producing herds were less carefully managed, including the calf-rearing unit.

18.1.2 Size of operation

Larger calf-rearing units had higher mortality rates. A large unit is defined as one rearing more than 30 calves over a 3-month period, which would equate to 120 calf/yr unit on a year-round calving herd. One could conclude that in the US, bigger units are less well managed than smaller units. This is not necessarily the situation in Asia and other less well-developed tropical dairy regions.

18.1.3 First colostrum feeding management

Newborn calves allowed to suckle their dam for their first drink of colostrum have higher death rates than do those fed colostrum by hand. Removing the calf from its dam immediately after birth reduces the chances of ingestion of faecal material as the calf looks for the teat. Furthermore, a controlled feeding of sufficient, good-quality colostrum within a short period after birth will ensure absorption of sufficient colostral antibodies into the calf's bloodstream. Depending on nature to do this with modern day milking cows is a lot more haphazard.

18.1.4 Group size

Calves reared in groups of seven or more have higher death rates than calves reared in groups of six or less. The smaller the group size, the better individual attention for each calf.

18.1.5 Gender of rearer

Calves reared by men had higher death rates than calves reared by women. One can only conclude that women have better rearing skills than men, which may be related to their more developed maternal instinct. In my experience, the best calf rearers are female nurses, because they have been trained to anticipate health problems before they happen.

18.1.6 Relationship of rearer to farm owner

Calves reared by the farmer's children or employees have higher death rates than calves reared by the farm owner or spouse. One can conclude that if you own the calves, you will take more care with them.

18.1.7 Time of feeding roughage

Higher death rates were found on farms where feeding of hay or other roughages was delayed until calves were 20 days or more old. The earlier that calves are offered roughage, the sooner their rumen begins to develop and the sooner they are likely to be

Best management practices provide practical guidelines for herd and feeding management of milk-fed calves.

weaned, either voluntarily or because of the feeding system. There was no effect of age of feeding concentrates or free choice of water on calf death rates.

18.1.8 Feeding mastitic or antibiotic milk

Heifer calves reared in units where mastitic or antibiotic milk was fed to them had higher death rates then in units where it is discarded, or fed to less valuable calves.

18.1.9 Feeding whole milk to calves

Calves fed whole milk from a bulk tank had lower death rates than calves not fed whole milk. It is assumed that calves not fed whole milk were fed milk replacer.

These nine factors sum up a good calf-rearing system, in which calves are provided with close attention to their health, digestive development and welfare by a person who really cares for them.

18.2 Monitoring your calf- and heifer-rearing system

It has been frequently stated that 'if you cannot measure it, you cannot manage it'. In the course of their operations, calf rearers already do collect, and can easily collect more, data on their stock. Much of this data could be, but probably has not been, used in making management decisions to improve the profitability and efficiency of their operation. Such decisions include:

- By recording the ear tag of any calf requiring veterinary treatment then recording how many lactations it remains in the milking herd, decisions could be made on whether similarly treated calves should be sold or still kept as replacements for the milking herd.
- Once farmers know the total costs for their first-calf heifers to enter their dairy herd and start generating income, they can then decide on whether it is more profitable to sell all their calves and rely on purchased in-calf heifers to maintain or expand herd sizes.
- By monitoring feeding and management costs from weaning to first calving, farmers can compare that with purchasing pregnant heifers reared on other farms.
- By monitoring live weights and wastage rates, at different stages of rearing, farmers can decide on optimum target live weights, and hence feeding management, for their particular operation.
- In many Western countries, milk processors require farms to enter audits for quality assurance programs. If this eventually occurs in tropical Asia, some form of record keeping will become mandatory.
- Record keeping can assist with identifying areas requiring attention and to direct staff to problem areas or potential risk areas on farm.

The following sections list some of these measures that can be easily collected and used in making these future decisions.

18.2.1 Pre-calving (heifer's dams)

- genetic merit of dams of replacement heifers
- costs of semen and hence each live heifer calf.

18.2.2 Post-calving (heifer's dams)

- calving difficulties (%)
- calves born dead (%)
- colostrum quality (proportion of different quality categories)
- average time after birth that colostrum is administered to newborn calves
- quantity of colostrum fed plus subsequent calf antibody status.

18.2.3 Pre-weaning

- litres of vat milk used to feed each calf
- weekly concentrate intake: as a guide to weaning age
- proportion of calves that die, are sick, or sold (and why) (%)
- record ear tag of each treated calf: to assist in future decision making about their fate
- average weaning age
- approximate time spent on rearing calves (hr/day and hence min/calf until weaning)
- costs of purchased feeds
- costs of veterinary treatment (drugs and visits)
- costs for routine management (vaccines, drenches, etc.)

- capital cost of shed and equipment: to calculate costs for depreciation
- live weight and wither height at 12 weeks of age: to compare pre-weaning performance from year to year.

18.2.4 Pre-mating

- weekly concentrate and hay inputs: to help plan future feeding programs as they vary with season
- quality of supplements
- live weight and wither heights at 6, 9 and 12 months, and at mating at 15 months
- conception rate at mating
- inseminations per conception, if using artificial insemination
- faecal egg counts at strategic times: to assist with drenching program
- proportion of heifers that die, are sick, or sold (and why) (%)
- costs of purchased feeds, veterinary treatment and routine management
- costs of mating (semen and oestrus synchronisation, or bull)
- total rearing costs per calving heifer: the 'bottom line'.

18.2.5 Post-calving

- days to successful insemination
- inseminations per conception
- proportion of first lactation heifers that die, are sick, or sold (and why) (%)
- first lactation yield of milk or milk solids
- first lactation yield as a percentage of yields on mature cows
- wastage rate from birth to second calving, as a percentage of heifer calves reared.

18.3 What is best management practice and quality assurance?

Although heifer-rearing programs must be tailored towards individual producers, there are several general principles associated with all 'best management practices'. Those relevant to heifer-rearing programs, which producers should aim for are:

- incorporating heifer rearing into a business plan for the entire dairy enterprise
- making a commitment to continuous improvement in rearing costs, timeliness of each phase of the program and on the end product, namely heifer quality
- developing closer relationships and alliances with all outside service providers to the program, such as veterinarians, feed suppliers, dairy advisers, semen suppliers and AI technicians
- using performance recording, then benchmarking achievements with the industries' best practice for performance indicators such as heifer wastage rates, heifer milk production (as a proportion of herd average), heifer fertility and cost per first lactation heifer
- integrating environmental and animal welfare concerns in all aspects of the program
- being involved with other producers to improve knowledge and upgrade the competitiveness of the dairy enterprise.

Best management practices provide practical guidelines for herd and feeding weaned heifer management.

The six lists of questions in the following sections were developed for a quality assurance program developed in the US (Dairy Quality Assurance Center 1998). The major objective of the program is to increase herd performance while at the same time reducing herd costs and meeting customer demands for animal care and quality animal products. These checklists document the most appropriate questions dairy farmers, their staff, advisers and suppliers of services to their heifer-rearing program should seek to answer. Each question provides for a 'yes' or 'no' answer. Just because farmers and advisers tick the 'no' box does not necessarily indicate that they are not doing a good job with their young stock management. It could simply indicate that they have considered the question and decided it is not relevant to their particular heifer management program.

This audit was developed on accepted norms for US dairy producers. However, quite a few of these practices have yet to be accepted as routine by Asian producers, such as testing all cows for certain diseases. Nevertheless, such practices have been included in the following checklists, because in the future, they may become integrated into Asian dairy production systems.

18.3.1 Planning general herd and heifer management

Profit starts and ends with a commitment by owners and/or managers of dairy enterprises to seek quality. Unless management is committed to improve quality, few gains will be achieved by producers and service providers associated with heifer

Table 18.1. A checklist for general herd and heifer management

Yes	No	Best management practices checklist
		Do you consistently meet milk quality standards for somatic cell counts and bacterial levels?
		Do you believe it is more profitable to increase milk quality and yield via replacement heifers than to improve milk quality and yield from the current herd?
		Do you have a permanent 'tamper-proof' animal identification system in place?
		Can you readily track and validate to others the quality represented in your replacement heifers?
		If you rear your heifers off farm, do you have in place a measurement system to evaluate such a rearing program?
		Have you been able to consistently produce more milk per cow each year?
		After accounting for purchased supplements, have you been able to consistently increase milk production per hectare of forage grown on farm each year?
		Do your first lactation heifers consistently produce >80% of the full lactation milk yield of your mature cows?
		Is your target live weight at first calving based on breed and target milk yield of your mature cows in the herd?
		Do you participate in other quality assurance audits in a quality assurance program?

operations. A commitment to producing quality replacement heifers is one important step towards increasing herd health and quality milk production.

Producers should review the checklist in Table 18.1 to see how many of the practices they can tick in the 'yes' column.

18.3.2 Planning heifer supply programs

Quality replacement heifers programs can be described as those that produce strong healthy pregnant heifers at 24 months of age, after which the heifers become productive members of the milking herd for at least five lactations. Planning quality heifers starts with the present herd. The key time to start 'quality' replacement heifer programs is before mating.

Producers should review the checklist in Table 18.2 to see how many of the boxes they can tick in the 'yes' column.

Table 18.2. A checklist for planning heifer supply programs

Yes	No	Best management practices checklist
		Do you participate in a herd-testing program to help identify the best cows from which to produce your replacement heifer calves?
		Have you and your veterinarian developed a routine herd health program including appropriate vaccination schedules for your cows?
		Do you routinely score the body condition of your cows to evaluate management and assure that they are in good condition to produce thrifty, healthy calves?
		If seasonally calving, do you have a compact calving program with all your replacement heifers born within 6–8 weeks?
		If not, are you able to achieve this by using AI over your maiden heifers?
		Do you select sires that will breed top-quality replacement heifers?

18.3.3 Planning heifer care from birth to weaning

Quality replacement heifers start by being strong calves at birth, followed by a quality meal of colostrum and then consuming some solid feed within the first 2 weeks of age. Producers should review the checklist in Table 18.3 to see how many of the boxes they can tick in the 'yes' column.

Table 18.3. A checklist for planning heifer care from birth to weaning

Yes	No	Best management practices checklist
		Are your calf mortality losses less than 5% of calves born, prior to and during the first 24 hr of life?
		Are your calf mortality losses from 24 hr of age to weaning less than 3% of live calves born?
		Are your springing cows and heifers provided with a clean, dry area for calving?
		Do you remove the calf from her mother, preferably at birth, but at least within the first 12 hr?
		Do you dip the navel in a strong (7%) iodine solution immediately after birth?
		Do you have a good program to supply high-quality colostrum, such as: 1. Use a colostrometer to monitor colostrum quality and only feed good-quality colostrum? 2. Ensure the calf gets 2 or more litres immediately after birth? 3. Provide another 2 or more litres within 6–12 hr of birth? 4. Use a stomach tube, if necessary? 5. Clean bottles, buckets and equipment regularly? 6. Pool colostrum from older cows tested negative for Johne's disease and enzootic bovine leucosis?
		Do you test the blood of some calves to check the efficiency of your colostrum program?
		Do you remove calves to a clean, dry area, preferably at birth, but at least within the first 12 hr?
		Do you use a permanent form of identification for each calf?
		Do you minimise contact between batches of calves until about 5 weeks of age?
		Do you provide calves with access to water at all times?
		Do you provide at least 1–2 m^2 per calf during milk rearing?
		Do you provide concentrates to each calf within the first week of age?
		Do you ensure calves each consume 0.75 kg of calf concentrate for 2 or more days prior to weaning
		Do your calves weigh 100 kg (Friesian) or 90 kg (Jersey) by 12 weeks of age?
		Do you weigh calves using cattle scales?
		Do you monitor the health and welfare of calves at least twice each day?
		Have you discussed your calf health management program with your veterinarian?
		To minimise the spread of Johne's disease, do you remove all possible avenues of infection between adult animals and calves?
		Do you isolate any calves showing signs of ill health to minimise spread of infection, then feed them last?
		Do you know how much money can be saved by early weaning?
		Do you quarantine any pre-weaned calves introduced onto your farm?
		Do you use the best possible feeds in your program (milk replacer, concentrates, straw)?

Table 18.3. A checklist for planning heifer care from birth to weaning (continued)

Yes	No	Best management practices checklist
		Have you discussed the use of waste (antibiotic/mastitic) milk with your veterinarian?
		Do you dehorn and remove extra teats from heifer calves during the milk-rearing stage?
		When selling excess calves, do you follow all the regulations regarding suitability for sale (minimum age, health status, antibacterial residues) and transport them in a suitable trailer?
		Score each of the following six disease problems you may encounter during milk rearing. Use a ranking of 1 (least problem) to 6 (biggest problem). Scours or diarrhoea Respiratory problems Joint or navel problem Trauma Unknown Rarely have illness

18.3.4 Planning heifer care from weaning to mating

It is a challenge to produce quality heifers between weaning and mating. Once they have been weaned off milk, their feeding management is frequently neglected, yet this is one of the most important periods of their life. Heifers achieve puberty at about one half their mature size. The sooner they reach puberty and start cycling, the more likely they will conceive when mated at 15 months of age.

Producers should review the checklist in Table 18.4 to see how many boxes they can tick in the 'yes' column.

Table 18.4. A checklist for planning heifer care from weaning to mating

Yes	No	Best management practices checklist
		Do you have feeding strategies to minimise the growth check immediately after weaning?
		Does your focus on heifer growth include nutrition, health, parasite control and social factors?
		Do you understand the importance of achieving target minimum live weights and wither heights for age at every stage of heifer development?
		Do you routinely monitor heifer growth using cattle scales?
		Do you routinely monitor wither height?
		Do you understand the change heifers go through as they switch from a milk-based diet to a fully developed ruminant using solid feeds?
		Do you feed calf concentrates before, during and after weaning?
		Do you base your feeding program on growth rates, which can vary dramatically with the availability and quality of pastures being grazed?
		In year-round calving herds, do you group heifers on age and/or live weight?
		If growth rates fall below acceptable targets, do you supplement heifers with quality feeds, such as cereal grain and/or good-quality conserved forages?

Table 18.4. A checklist for planning heifer care from weaning to mating (continued)

Yes	No	Best management practices checklist
		Are you aware of potential problems (fatty udder syndrome) arising from feeding excess high-energy/low-protein feeds between weaning and puberty?
		To minimise the spread of Johne's disease, do you ensure heifers under 12 months of age do not graze pastures that have been stocked with adult animals during the previous 12 months?
		Have you developed a health management program (vaccinations, internal and external parasite control) in conjunction with your veterinarian?
		Do you use individual needles during any vaccination program requiring intramuscular injections, disinfecting needles in alcohol?
		If your herd is diagnosed with enzootic bovine leucosis, are you meticulous in ensuring no cross contamination of animals with blood or milk?
		Do you have an effective fly control program, if necessary?

18.3.5 Planning heifer mating programs

Successful mating programs for replacement heifers require all animals to be cycling. All the hard work and quality management will only return profit to your operation if conception rates are high when heifers are mated at 15 months of age.

Table 18.5. A checklist for planning heifer mating programs

Yes	No	Best management practices checklist
		Do your replacement heifers have body condition scores of 5 to 6 units and are they gaining weight at mating?
		By 15 months of age, have all your replacement heifers achieved minimum target live weights for mating (330 kg for Friesians and 245 kg for Jerseys)?
		Do you treat your heifers for internal and, if required, external parasites just prior to mating?
		With seasonal calving herds, do you plan heifer calving dates in relation to those of your mature cows? This may be a week or two earlier to assist with feeding management of newly calved heifers.
		Are you aware of the benefits of using AI over natural mating?
		Are you aware of the benefits of using dairy (as against beef) bulls or semen?
		Do you use AI for mating well-grown heifers, then follow on with good-quality herd bulls to 'clean up' these heifers and any smaller ones not inseminated?
		If using AI, do you select semen from sires or breeds selected for ease of calving?
		Have you selected the most appropriate heat-detection procedure for your operation?
		Do you choose to use heat synchronisation, if appropriate, for your heifer mating program?
		If using natural mating, do you select sufficient good-quality bulls, taking note of their mobility and libido (one bull per 30 heifers plus one spare)?
		If using natural mating, do you ensure all bulls have been vaccinated against vibriosis?
		Do you routinely test your heifers for pregnancy to plan their calving program?

Producers should review the checklist in Table 18.5 to see how many of the boxes they can tick in the 'yes' column.

18.3.6 Planning heifer care from mating to calving

Good heifer management is vital up to the point of calving, particularly if the calf is destined for the replacement herd.

Producers should review the checklist in Table 18.6 to see how many of the boxes they can tick in the 'yes' column.

Table 18.6. A checklist for planning heifer care from mating to calving

Yes	No	Best management practices checklist
		Do your heifers gain on average 0.6–0.8 kg/day after mating?
		Do your heifers calve down in body condition score 5 to 6 units?
		Do you introduce your heifers to the milking shed (or at least run them through the milking shed) prior to calving?
		Do you avoid mixing replacement heifers with older dry cows?
		Do you avoid high somatic cell counts by keeping replacement heifers in a clean dry paddock for at least 1 month before calving?
		Do you store good-quality, tested colostrum (either freshly chilled or frozen from the previous year) from older cows to feed all calves routinely from first-calf heifers?
		Have you and your veterinarian developed a health treatment program for heifers pre- and immediately post-calving?

18.4 Ensuring the relevance of these BMPs to tropical small holder dairy farmers

I believe that the above collection of checklists can form the basis of a quality assurance scheme for rearing dairy replacement heifers in tropical Asia. However, it must be emphasised that this set of BMPs was specifically developed for a modern dairy farm operating in the US (or other Western countries) where farmers can find technical support from a variety of sources. This is rarely the case in Asia because farms are generally very small, farmers have limited technical support and, in many cases, they may not even have the technical skills to decide if they are doing a good job or not.

That is not to say the lists are not really relevant to SHD farmers in tropical Asia. They can become an integral part of any training program for farmers that is conducted by government or dairy cooperative advisers. Furthermore, they can become incorporated into the curriculum for a university undergraduate program to educate our future dairy advisers, research scientists and other dairy stakeholders.

As they stand now, some cannot easily be acted upon. However, it is up to the users of the technical information contained in this book to select those most appropriate for their training programs and perhaps rephrase or reinterpret them to make them more relevant for their students or program participants. It is better for the trainer or educator to place them into context for their particular target audience than for them to be taken directly as they are presented in the tables above.

Appendix 1: John Moran's golden rules of calf and heifer rearing

John Moran's ten golden rules of calf rearing

- Ensure each calf receives 4 L of top-quality colostrum within 6 hr of birth. Remember the 3 Qs for colostrum feeding (quality, quantity, quickly). Dip or spray the umbilical cord with iodine solution.
- Remember that feeding milk only once each day encourages faster rumen development, reduces rearing costs, ensuring fewer health problems and better post-weaning performance.
- Provide continual access to clean water and high-quality concentrates from day 1. Also provide a palatable roughage source, such as clean straw.
- Give individual attention to each calf and make time to check at least twice daily for signs of ill-thrift or sickness.
- Develop a disease action plan that includes good hygiene, isolation of sick calves, fluid replacement and TLC (tender loving care). Drugs should only be used a last resort, to complement a well-managed system.
- Milk rear calves in a clean, dry, well-ventilated shed, in groups no more than six animals, providing at least 1.5 m^2 per calf.
- Ensure good record keeping, because this will help pinpoint problems in your system.
- Minimise stresses by following set routines each day, reducing over crowding and 'keeping your troubles' out of the calf shed.
- People rear calves, systems don't! Do not rear calves if you don't enjoy it – find a specialist.
- A good calf-rearing system produces healthy, fully weaned calves weighing 90–100 kg at 12 weeks of age.

John Moran's golden rules of heifer rearing

The principles of good heifer rearing can be summarised in the following key points:

Targets

- Ranges of target live weights for ages with Friesian and Jersey heifers in well-managed herds are:

Age (months)	Friesian	Jersey
3 (fully weaned)	90–110	65–85
6	150–175	110–130
9	210–235	155–180
12 (yearling)	270–300	200–230
15 (mating)	330–360	245–275
18	390–420	290–320
21	455–485	335–365
24 (pre-calving)	520–550	380–410

- The optimum pre-calving live weight of heifers varies with their target milk yield as mature cows. In Friesians, this can range from 500 kg in herds averaging 5000 L/cow/yr to 560 kg in herds averaging 7000 L/cow/yr.
- During their first lactation, well-reared heifers should produce at least 80% of the full lactation milk yield of their mature herd mates.
- Heavier heifers must be fed well to achieve their economic benefits. There is little point in growing out heavy heifers then underfeeding them as milkers.
- Heifers should be managed to grow at an average of 0.7 kg/day from weaning to first calving, although this can vary during the 24 months from 0.5 to 1.0 kg/day depending on seasonal conditions.

Feeding

- Heifers should be provided with a good-quality diet for their first 12 months, containing 10–11 MJ/kg DM of energy and 12–16% protein.
- High-energy supplements are often required to achieve target growth rates of young heifers (up to 6 months of age), particularly during their first winter.
- Any hay or silage fed must be of good quality: at least 10 MJ/kg DM of energy and 14% protein.
- Be wary of feeding an unbalanced diet containing too much low-protein grain during the 3–9 months of age critical period, because excessive growth rates can lead to fatty udders and reduced milk potential.

Management

- If young stock are allowed to lose weight or grow very slowly for lengthy periods, they will not achieve their potential frame size.
- Low mating live weights can lead to calving difficulties 9 months later. Excessive feeding after mating can also result in dystocia. Dystocia reduces milk yield and increases the number of days to the second conception.
- Use artificial insemination and quality Friesian semen on well-grown heifers to provide replacement calves from first-calf heifers.
- Heifers should be regularly weighed, at least every 3 months, with wither heights recorded at each weighing.

Appendix 2: Conversion of units of measurements

1. Abbreviations

k	kilo or thousands
M	mega or millions
mm	millimetre
cm	centimetre
m	metre
ha	hectare
mL	millilitre
L	litre
J	joule
MJ	megajoule
min	minute
hr	hour
yr	year
mg	milligram
g	gram
kg	kilogram
t	tonne
lb	pound
ft	foot
$	dollar
c	cent
<	less than
>	greater than

2. Conversion of Imperial units to metric units

Length:	1 inch = 25.4 mm
	1 foot = 30.5 cm

	1 yard = 0.91 m
	1 mile = 1.61 km
Volume:	1 cu ft = 0.028 m^3
	1 pint = 0.57 L
	1 gallon = 4.54 L
	1 bushel = 36.4 L
	1 acre foot = 1.23 ML
Area:	1 acre = 0.40 ha
	1 sq mile = 2.59 sq km
Weight:	1 ounce = 28.3 g
	1 pound = 0.454 kg
	1 hundred weight = 50.8 kg
	1 long ton = 1017 kg (2240 lb)
Energy:	1 calorie = 4.19 joules
Density:	1 lb/ft^3 = 0.063 kg/m^3
Rate:	1 gallon/acre = 11.23 l/ha
	1 pound/acre = 1.12 kg/ha
	1 gallon/ton = 4.17 l/tonne
Pressure:	1 pound/sq in (psi) = 1.45 kPa (kilopascals)
Yield:	1 lb/ac = 1.12 kg/ha
Temperature:	1°F = ((9/5) × C) + 32
	1 degree F is equivalent to 0.56 degrees C
	50°F = 10.0°C
	60°F = 15.6°C
	70°F = 21.1°C
	80°F = 26.7°C
	90°F = 32.2°C
	100°F = 37.8°C
	110°F = 43.3°C

3. Conversion of US units to metric units

Volume:	1 gallon = 3.79 L
	1 bushel = 35.2 L
Weight:	1 hundred weight = 45.4 kg
	1 short ton = 907 kg (2000 lb)
Milk prices:	\$10/hundred weight = 22.0 c/L

Forage maize yields @ 30% DM:

	25 ton fresh weight/acre = 16.8 t DM/ha
Food energy:	1% unit TDN = 0.185 MJ/kg DM of metabolisable energy
	30% TDN = 3.7 MJ/kg DM of ME
	40% TDN = 5.5 MJ/kg DM of ME
	50% TDN = 6.4 MJ/kg DM of ME
	60% TDN = 7.4 MJ/kg DM of ME

70% TDN = 8.3 MJ/kg DM of ME
80% TDN = 9.2 MJ/kg DM of ME
1 MCal/lb = 9.22 MJ/kg
1 MCal/kg = 4.19 MJ/kg

4. Conversion of other specific country units to metric units

Most countries now use the metric units of measurement, but certain countries have their own historical units, which are still used by farmers and advisers.

China

Length:	1 chi = 33 cm
	1 li = 500 m
Volume:	1 gongsheng = 1 L
Weight:	1 jin = 500 g

Thailand

Length:	1 nui = 2.1 cm
	1 kheup = 25 cm
	1 sawk = 50 cm
	1 waa = 2 m
	1 sen = 40 cm
	1 yoht = 16 km
Weight:	1 baht = 15 g
	1 tamleung = 60 g
	1 chang = 1.2 kg
	1 haap = 60 kg
Area:	1 sq waa = 4 sq m
	1 ngaan = 400 sq m
	1 rai = 1.6 ha

Appendix 3: Currency converter for South and East Asia

Instead of expressing costs and returns in one currency (conventionally US dollars), this manual makes use of currencies from various South and East Asian countries. For the reader's benefit, rather than convert them all to a single currency in the text, the following currency converter can be used to compare their values in January 2012. More up-to-date conversions can be obtained via the internet from a currency converter located at <www.xe.com/ucc/>.

	AFN	AUD	BDT	CNY	INR	IDR	MYR	PKR	PHP	LKR	THB	USD	VND
AFN	X	49.6	0.589	7.66	0.91	5.31	15.3	0.534	1.09	0.424	1.52	48.3	2.30
AUD	0.020	X	0.012	0.153	0.018	0.011	0.309	0.011	0.022	0.009	0.031	1.035	0.046
BDT	1.70	84.2	X	13.00	1.54	90.90	26.0	0.907	1.86	0.553	2.59	69.50	3.90
CNY	0.130	6.52	0.077	X	0.119	0.069	2.00	0.070	0.144	0.047	0.200	6.30	0.299
INR	1.09	55.0	0.645	8.435	X	5.08	16.8	0.586	1.213	0.47	1.689	53.14	2.52
IDR	1.88	9.50	1.11	1.457	1.72	X	28.9	1.01	2.095	0.80	2.918	91.79	0.043
MYR	0.065	3.25	0.038	0.489	0.059	0.035	X	0.035	0.072	0.028	0.100	3.139	0.149
PKR	1.87	93.3	1.102	14.37	1.706	9.91	28.7	X	2.058	0.79	2.865	90.15	4.430
PHP	0.91	45.4	0.537	6.89	0.830	4.83	13.98	0.486	X	0.39	1.39	48.31	2.01
LKR	2.36	117.9	1.139	18.06	2.147	1.25	36.2	1.26	2.600	X	3.62	113.9	5.41
THB	0.655	32.6	0.386	5.018	0.597	0.347	10.00	0.35	0.718	0.260	X	32.61	1.50
USD	0.021	0.966	0.012	0.158	0.018	0.011	0.317	0.011	0.022	0.009	0.031	X	0.047
VND	0.435	21.59	0.256	3.334	0.397	0.231	6.67	0.232	0.478	0.184	0.66	21.1	X

Abbreviations

AFN: Afghanistan afghanis
AUD: Australian dollar
BDT: Bangladesh taka
CNY: China yuan renminbi
INP: India rupee
IDR: Indonesian rupiah × 100
MYR: Malaysian ringgit
PKP: Pakistan rupee
PHP: Philippines peso
LKR: Sri Lanka rupee
THB: Thai baht
USD: US dollar
VND: Vietnam dong × 1000

Appendix 4: Workshop expectation and evaluation forms

IMPROVED YOUNG STOCK MANAGEMENT ON TROPICAL DAIRY FARMS

Expectations of workshop

Date/Location: ....

1. Name: ....
2. Address: ....
3. Position held (farmer, dairy cooperative staff, milk collection centre staff, government adviser):

   ....

4. How many milking cows do you have?

   ....

5. What is your total number of dairy stock (calves, heifers, cows, bulls)?

   ....

6. How many litres of milk each day do **all** your milking cows produce (on average)?

   ....

7. How many dairy heifer and bull calves do you rear each year?

   ....

8. How many of these heifer and bull calves die each year?

   ....

9. What topics would you like to learn about in this workshop?

   a) ....

   b) ....

   c) ....

Please answer the following questions with a Yes or No.

	Yes/No
10. Do you have a good calf-rearing system?	
11. Do you have a good system for growing out your weaned heifers?	
12. Do you think farmers can improve their calf- and heifer-rearing systems?	
13. Do you think government advisers can help you to improve your calf- and heifer-rearing system?	

IMPROVED YOUNG STOCK MANAGEMENT ON TROPICAL DAIRY FARMS

Evaluation of workshop

Date/location: ....

Participant's name: ....

1. **Expectations:**
   What were your expectations of the workshop? Please list:

2. **Outcome:**
   What knowledge have you gained from this workshop?

3. **Relevance of training:**
   Please describe how this training will be of use to your work.

4. **Program delivery:**
   Please tick the appropriate space to indicate your views on the way the workshop has been delivered.

	Not enough	About right	Too much
Overall program			
Lectures and/or formal instruction			
Discussion			
Visits on site/fieldwork			
Reading matter provided			

5. **Services:**
   How do you rate the services provided for you? (Please tick)

	Excellent	Good	Fair	Not good
Training/trainers				
Training location				
Other				

6. **Other comments:** ....

   ....

7. **Overall assessment:**
   How do you rate this program in terms of its relevance to your role in the dairy industry? (Please tick)

	Excellent	Good	Fair	Not good
Personal relevance to you				

8. **What are the weaknesses of the workshop?**

....

....

....

9. **What improvements can be made for future workshops?**

....

....

....

10. **List the most important messages/information that you found most useful to you.**

....

....

....

11. **List the least useful messages/information that you found least useful to you.**

....

....

....

Please rank the following questions for their importance to you (1 to 5), where 1 is low/not much and 5 is high/a lot.

	Score
12. How do you rate farm visit?	
13. How do you rate small groups and reporting back sessions?	
14. How do you rate overhead presentations?	
15. How do you rate importance of improved young stock management skills in your job?	
16. How much have you improved your knowledge of young stock management skills?	
17. How well will you be able to apply knowledge to farmer situations?	
18. When should you do a refresher course? Please score 1 for 3 months; 2 for 12 months; 3 for 2 years; 4 for never	

Thank you for your participation in this workshop

Glossary

Terms in this glossary are defined in the context of their use in this book.

abomasum: the fourth (or true) stomach in ruminant animals where feeds are digested using enzymes produced in the stomach wall.

acid detergent fibre (ADF): the less digestible or indigestible parts of the fibre; that is, the cellulose and lignin only.

acidosis: an excessive increase in rumen acid caused by feeding too much grain or other starchy feeds or by introducing them into the diet too quickly.

***ad lib* or *ad libitum*:** fed to appetite.

age at first calving (AFC): a good indicator of heifer management in year-round calving herds. In seasonal-calving herds, it is usually pre-determined at about 24 months of age.

AI: artificial insemination.

antibiotics: drugs, generally prescribed by veterinarians, that treat diseases by killing specific bacteria. Unfortunately, their use is becoming too prevalent in normal calf rearing, such as their inclusion in commercial milk replacers.

antibiotic residues: antibiotics remaining in animal products when sold for human consumption.

antibodies: protective proteins produced by animals in response to specific diseases. These are passed on to newborn calves through the colostrum.

Australian Milking Zebu (AMZ): a tropically adapted dairy breed developed in Australia based on a Jersey and Red Sindhi cross.

best management practice (BMP): a description of the most suitable procedures for undertaking a set of tasks and is the basis used to develop a checklist for planning various aspects of the calf and heifer operations. Basically it is 'saying what you should do, doing what you say, and then recording what you have done'.

biosecurity: the protocols introduced to minimise the introduction of diseases into the calf shed.

bloat: a condition caused by over distension of the abomasum or rumen, which requires immediate attention because it can quickly kill animals.

body condition: an estimation of stored body fat reserves. In Australian dairy cattle, it is estimated using a score from 1 (thin) to 8 (fat).

buffers: chemicals that prevent sudden changes in rumen pH.

Ca: calcium.

calving ease: a rating used when selecting dairy sires based on the estimated percentage of calves born to a particular sire that cause calving difficulties in mature cows.

calving interval (CI): the average time period between consecutive calvings in a dairy herd. The target is 12 months, although this is rarely achieved.

***Clostridia*:** bacteria causing a variety of diseases in calves and older cattle.

CMR: calf milk replacer.

***Coccidia*:** microbes, called protozoa, which cause scouring in calves.

colostrometer: a device that measures the level of immunoglobulins in colostrum. It is sometimes called a colostradoser.

colostrum (or beastings): the milk produced by cows for the first two milkings post-calving, which contains high levels of nutrients and immunoglobulins for transferring immunity onto newborn calves.

comfort zone: the range of air temperature when there is no measurable fluctuation in physiological processes of cattle.

conception rate: the proportion of the total number of services or inseminations that results in pregnancy.

conception to calving interval (CCI): the time period between when a cow has a calf and she next becomes pregnant.

contract heifer rearing: when dairy producers develop formal agreements with other graziers to grow out their heifers, usually at a pre-determined growth rate, until the point of calving.

controlled internal drug release (CIDR): intravaginal progesterone implants used to synchronise oestrus to better manage artificial insemination programs.

critical period: the period of a heifer's growth during which time the mammary tissue develops in the udder. During this period, which occurs around puberty, excessive growth or very low protein diets can lead to fatty tissue being deposited, instead of milk-producing tissues.

crude fibre (CF): a measure of fibre in the diet, now considered unacceptable because it does not always take into account all of the constituents that make up the fibre component of a feed; it measures only the alkali-soluble lignin and the cellulose.

crude protein (CP): a measure of the total protein in a feed, calculated as the total nitrogen content multiplied by 6.25. It includes true protein, which provides the amino acids for animal growth, and also non-protein nitrogen, such as urea.

***Cryptosporidia*:** microbes, called protozoa, which cause scouring in calves.

degradability: a measure of the degree of breakdown of dietary protein by rumen microbes.

dehorning: removal of small horns and horn buds in calves more than 2 months of age.

digestibility: the proportion of the dry matter in a feed that gets digested; it is the difference between what is eaten and what comes out as manure.

disbudding: removal of horn buds in calves younger than 2 months.

DPI: Department of Primary Industries (Victoria).

dry matter (DM): the proportion of a feed remaining after being dried at 80–100°C for 24 hr or until a constant dry weight is achieved. The nutritive value and the livestock requirements of feeds are usually expressed on a dry matter, rather than a fresh weight, basis.

duodenum: the first section of the small intestine.

dystocia: the technical term for calving difficulties, which are more common in first-calf heifers than in older cows.

***E. coli*:** bacteria that cause scours in calves.

electrolytes: mineral salts used to alter the pH of gut contents for optimum digestion. Electrolyte solution is a solution of salts (and often an energy source, such as glucose) used to replace fluids lost during scouring.

enzymes: chemicals produced by animals that assist with the breakdown of feeds in the digestive tract; examples are pepsin, lactase, rennin, lipase and galactase.

fatty udder syndrome: excess deposition of fatty tissue in the developing udder, which can reduce lifetime milk production. It can occur if heifers grow too fast (greater than 0.8 kg/day) prior to puberty, say between 3 and 10 months of age.

fibre: the cell wall, or structural material, in a plant. Fibre is made up of (among other things) cellulose, hemicellulose and lignin.

five in one: a vaccine used to protect against *Clostridia* bacteria.

flight zone: the 'personal' space around animals where they will attempt to move away from people.

heifer farms: a new initiative in many Asian countries in which calves are collected from individual farms and group reared in one location prior to their return to that farm just prior to calving down.

immunoglobulins (Ig): blood proteins (antibodies) in colostrum that pass on passive immunity to newborn calves.

Johne's disease: an incurable bacterial infection of the intestines. Heifers are highly susceptible up to 12 months of age, so must have no contact with mature animals or their faeces. This means that calves must be reared in complete isolation from milking cows and weaned heifers must only graze 'clean' paddocks for their first year of life.

joint-ill (or navel-ill): a bacterial infection of the umbilical cord in newborn calves that can cause arthritis of the joints.

key performance indicator (KPI): a numerical descriptor of some aspect of well-managed farm or herd performance that can be used as a realistic target for future improved farm management programs.

leptospirosis: a bacterial disease that is prevalent among dairy farmers due to its transmission from cows and calves.

live weight at first calving (LWFC): the live weight of heifers just before their first calving. In some instances, live weights may be recorded post-calving, in which case they will be about 80 kg lower than in-calf LWFC.

lower comfort zone temperature: the air temperature at which the energy intake must increase to minimise reduction in weight loss in growing cattle or to prevent weight loss in mature cattle.

medicine disease: a condition caused by prolonged use of antibiotics, which can upset the balance of rumen microbes.

metabolisable energy (ME): the amount of energy provided by a feed after deducting energy lost to faeces, urine, heat and gas production; it is the energy available to be used by the animal for its metabolic activities.

microbial protein: an important component of the microbes in the rumen, which is broken down into amino acids for use by the animal.

MJ ME/kg DM: megajoules of metabolisable energy per kilogram of dry matter.

MR: Malaysian ringgits.

N: nitrogen.

neonatal diarrhoea: the technical term for calf scours.

neutral detergent fibre (NDF): a measure of all the fibre (hemicellulose, lignin and cellulose) in a feed; it indicates how bulky the feed is.

non-protein nitrogen (NPN): not actually protein but simple nitrogen; however, microbes can make protein from simple nitrogen if enough energy (carbohydrates) is available in the rumen at the same time.

nurse cows: cows used for multiple suckling calves, either by running with them at pasture (continuous or foster suckling) or by holding them in specially designed races (restricted or race suckling).

oesophageal groove: a small channel in the rumen wall controlled by muscles, which allows liquid feeds to bypass the rumen for digestion directly in the abomasum.

P: phosphorus.

pancreas: an organ that produces enzymes to assist with digestion of milk products.

passive immunity: resistance against diseases passed from cow to calf via the immunoglobulins in colostrum. It is also called acquired immunity.

pellets: commercially produced and pelleted mixtures of feeds specially formulated for rearing calves or feeding specific types of livestock. They are generally based on cereal grains and other concentrates, but can also include agro-industrial by-products. They generally include specific mineral and vitamins.

pH: a measure of acidity or alkalinity on a scale from 1 (extremely acid) to 14 (extremely alkaline).

PKC: palm kernel cake.

probiotics: additives, usually bacteria, to improve the natural process of digestion.

pyloric sphincter: the valve at the end of the abomasum that controls the movement of feed into the duodenum.

(the 3) Qs: the targets for colostrum feeding management, namely quality, quantity and quickly.

quality: in relation to feeds, it is an indication of the level of energy and digestibility. In relation to milk, it refers to the level of various contaminants in milk, such as bacterial, chemical or any other adulterations that can be detected.

quality assurance (QA): a structured set of best management practices.

retained foetal membrane (RFM): membranes from newborn calves still inside the cow following birth.

rotavirus: a type of virus that cause scours in calves.

Rs: Sri Lankan rupees.

rumen: the major stomach in adult ruminants containing millions of microbes that breakdown feed particles prior to digestion by the animal. It is underdeveloped and non-functional in newborn calves.

rumen degradable protein (RDP): the portion of protein in the diet that is digested and used by the microbes in the rumen to build themselves, if enough energy (carbohydrates) is available at the same time.

rumen undegradable protein: *see* undegradable dietary protein.

***Salmonella*:** bacteria causing severe scouring in calves. These can also be transmitted to humans.

SHD: small holder dairy.

springing cow: a cow due to calf imminently.

standard operating procedures (SOP): a set of instructions for any activity or set of tasks undertaken on the farm.

starter: a name given to the first type of concentrate fed to calves during milk rearing.

submission rate: the proportion of the herd inseminated at least once in a given period of time (e.g. the first 10, 21, 24 or 30 days of mating).

supplement: a feed or product added to an animal's diet to increase the intake of some dietary component, such as energy, protein, fibre, vitamins or minerals.

target live weights and wither heights: pre-determined live weights or wither heights at certain ages, set as targets for growing heifers. Because they are targets for all heifers to achieve by a given age, they are minimums not average growth targets.

temperature humidity index (THI): a system for quantifying heat stress based on temperature and humidity. The higher the index, the greater the discomfort. This occurs at lower temperatures for higher humidities.

tender loving care (TLC): the term used for calf rearers with sufficient empathy towards calves to ensure they place close attention to their daily needs and management.

transition milk: milk from freshly calved cows (following colostrum) that milk processors will not collect for the first few days post-calving. It is usually (and incorrectly) referred to as colostrum.

undegradable dietary protein (UDP): dietary protein that escapes microbial digestion in the rumen and is broken down by the animal in the abomasum or duodenum.

VND: Vietnam dong.

wastage rate: a measure of losses in replacement heifers between birth and second calving.

WATCH: this summarises the key principles of good cleaning and sanitising of calf feeding equipment: namely water, action, time, chemicals and heat.

wither height: the height of heifers measured at the highest point on the shoulders, just above the front legs. Wither height is a good measure of bone growth in heifers, hence frame size.

withholding period: the number of days following drug administration before milk or meat can be sold from treated animals.

XB: crossbred.

zoonoses: calf diseases that can be passed onto humans.

References

Agricultural Research Council (1980). *The Nutrient Requirements of Farm Livestock, No. 2*, 2nd edn. Agricultural Research Council, London, UK.

Amuamuta, A., Asseged, B. and Goshu, G. (2006). Mortality analysis of Fogera calves and their Friesian crosses in Andassa Cattle Breeding and Improvement Ranch, Northwest Ethiopia. *Revue Med. Vet.* **157**, 525–529.

Anon (2010). Oestrus detection in cattle. The Dairy Site (website), May 2010. <http://www.thedairysite.com/articles/2362/estrus-detection-in-cattle>.

Asseged, B. and Birhanu, M. (2004). Survival analysis of calves and reproductive performance of cows in commercial dairy farms in and around Addis Ababa, Ethiopia. *Tropical Animal Health & Production* **36**, 663–672.

Bede, B. O. (2008). Dairy heifer rearing under increasing intensification of smallholder dairy systems in the Kenyan highlands. *Livestock Research for Rural Development* **20** (2) <http://www.lrrd.org/lrrd20/2/bedea20022.htm>.

Bebe, B. O., Abdulrazak, S. A., Ogore, P. O., Ondiek, J. O. and Fujihara, T. (2001). A note on risk factors for calf mortality in large-scale dairy farms in the tropics: A case study on Rift Valley area of Kenya. *Asian-Australasian Journal of Animal Science* **14** (6), 855–857.

Bebe, B. O., Udo, H. M., Rowlands, G. J. and Thorpe, W. (2003). Smallholder dairy systems in the Kenya highlands: cattle population dynamics under increasing intensification. *Livestock Production Science* **82**, 211–221.

BAMN (Bovine Alliance on Management and Nutrition) (1995). *A Guide to Colostrum and Colostrum Management for Dairy Calves*. BAMN, Arlington, Virginia, USA.

BAMN (1997). *A Guide to Modern Calf Milk Replacers. Types, Use and Quality*. BAMN, Arlington, Virginia, USA.

BAMN (2007). *Heifer Growth and Economics: Target Growth*. BAMN, Arlington, Virginia, USA.

Dairy Australia (2011). Rearing healthy calves. How to raise calves that thrive. Dairy Australia (website), Melbourne. <http://www.dairyaustralia.com.au/Animals-feed-and-environment/Animal-welfare/Calf-welfare/Rearing-healthy-calves-manual.aspx)>.

Dairy Quality Assurance Center (1998). *Quality Replacement Heifers: Growing Your Own Profits*. DQAC, Iowa, USA.

Davis, C. L. and Drackley, J. K. (1998). *The Development, Nutrition and Management of the Young Calf*. Iowa State University Press, Ames, Iowa, USA.

DCHA (Dairy Calf & Heifer Association) (2009). DCHA gold standards – Part V. Mortality & morbidity, *Tip of the week*, Sep 22, 2009. DCHA, Chesterfield , MO, USA.

DCHA (2010). Gold standards II, DCHA, USA <http://www.calfandheifer.org/?page=GoldStandardsII>.

De Jong, R. (1996). Dairy stock development and milk production with smallholders. Ph.D thesis. Wageningen University, The Netherlands. <http://library.wur.nl/WebQuery/wda/abstract/929615>.

Fisher, J. (2009). Standard Operating Procedures vital. *Australian Dairy Farmer* **Nov/Dec 2009**, 85.

Fowler, M. (1999). What is it worth to know a calf's Ig level? *Proceedings of the 3rd Professional Dairy Heifer Growers Association Conference*, pp. 31–36, Bloomington, Minnesota.

French, N. P., Tyrer, J. and Hirst, W. M. (2001). Smallholder dairy farming in the Chikwaka communal land, Zimbabwe: birth, death and demographic trends. *Preventive Veterinary Medicine* **48**, 101–112.

Gitau, G. K., McDermott, J. J., Walter-Toews, D., Lissemore, K. D., Osumo, J. M. and Muriuki, D. (1994). Factors influencing calf morbidity and mortality in smallholder dairy farms in Kiambu district of Kenya. *Preventive Veterinary Medicine* **21**, 167–177.

Heinrichs, A. J. (2002). *CalfTrack: A Calf Management Training System.* Pennsylvania State University, USA.

Heinrichs, A. J. and Jones, C. M. (2002). *Feeding the Newborn Dairy Calf.* Pennsylvania State University, USA.

Heinrichs, A. J. and Swartz, L. A. (1990). *Management of Dairy Heifers.* Extension Circular 385. Pennsylvania State University, USA.

Humphris, T. (1998). The effect of natural suckling, compared with natural suckling and artificial supplementation with colostrum, on the level of passive transfer of immunoglobulins in calves. *Proceedings of the XXth World. Association for Buiatrics Congress*, Sydney. pp. 345–50.

Ibrahim, M. N. (1988). *Feeding Tables: A Practical Guide.* p. 71. E.T.I. Division, Animal Production and Health, Kandy, Sri Lanka.

Jemberu, W. T. (2004). Calf morbidity and mortality in dairy farms in Dedre Zeit and its environs, Ethiopia. M. Vet. Sci. thesis, University of Addis Ababa, Ethiopia.

Kivaria, F. M., Noordhuizen, J. P. and Kapaga, A. M. (2006). Prospects and constraints of small holder dairy husbandry in Dar es Salaam region, Tanzania. *Outlook on Agriculture* **35** (3), 209–215.

Knopf, L., Komoin-Oka, C., Betschart, B., Gottstein, B. and Zinsstag, J. (2004). Production and health parameters of N'Dama village cattle in relation to parasitism in the Guinea savannah of Cote D'Ivoire. *Revue d'élevage et de Médecine Vétérinaire des pays Tropicaux* **57** (1-2), 95–100.

Krishnamoorthy, U. and Moran, J. (2011). *Rearing Young Ruminants on Milk Replacers and Starter Feeds.* FAO Animal Production and Health Manual No. 13. FAO, Rome. <http://www.fao.org/docrep/014/i2439e/i2439e00.pdf>.

Lanyasunya, T. P., Rong, W. H., Abdulrazak, S. A. and Mukisira, E. A. (2006). Effect of supplementation on performance of calves on small holder dairy farms in Bahati Division of Nakura District, Kenya. *Pakistan Journal of Nutrition*, **5** (2), 141–146.

Leadley, S. and Sojda, P. (2001). Improving heifer handling (Parts 1 and 2). *Calving Ease Newsletter*, Dec 2001/Jan 2002.

Lobago, F., Bekana, M., Gustafsson, H. and Kindahl, H. (2006). Reproductive performances of dairy cows in smallholder production system in Selalle, Central Ethiopia. *Tropical Animal Health Production* **38**, 333–342.

Losinger, W. and Heinrichs, A. (1997). Management practices associated with high mortality among preweaned dairy heifers. *Journal of Dairy Research* **64**, 1–11.

Lyimo, H. L., Mtenga, L. A., Kimambo, A. E., Hvelplund, T., Laswai, G. H. and Weisbjerg, M. R. (2004). A survey on calf rearing systems, problems and improvement options available for the small holder dairy farmers on Turiani in Tanzania. *Livestock Research for Rural Development* **16** (4) <http://www.cipav.org.co/lrrd/lrrd16/4/lyim16023.htm>.

Madalena, F. E., Teodoro, R. L., Lemos, A. M. and Barbosa, R. T. (1996). Comparative performance of six Holstein-Friesian x Guzera crossbred groups in Brazil. 8. Calf mortality. *Revista Brasileira De Genetica* **18**, 215–220.

Margerison, J. and Downey, N. (2005). Guidelines for optimal dairy heifer rearing and herd performance. In *Calf and Heifer Rearing: Principles of Rearing the Modern Dairy Calf from Calf to Calving.* (Ed P. C. Garnsworthy). pp. 307–338. Nottingham University Press, Nottingham, UK.

McNeil, J. (2009). Program to maintain calf rearing standards. *Australian Dairy Farmer* **Sep/Oct 2009**, 95–98.

Millar, C. (2010). Sexed semen, will it work for you? *Australian Dairy Farmer* **May/June 2010**, 18.

Menjo, D. K., Bebe, B. O., Okeyo, A. M. and Ojango, J. M. K. (2009). Analysis of early survival of Holstein-Friesian heifers of diverse sire origins on commercial dairy farms in Kenya. *Tropical Animal Health Production* **41**, 171–181.

Moran, J. B. (2002). *Calf Rearing: A Practical Guide.* 2nd edn. Landlinks Press, Melbourne.

Moran, J. B. (2005). *Tropical Dairy Farming: Feeding Management for Small Holder Dairy Farmers in the Humid Tropics.* Landlinks Press, Melbourne. <http://www.publish.csiro.au/nid/197/issue/3363.htm>.

Moran, J. (2006). The role of automatic calf feeders in Australian calf rearing systems. *Agricultural Science (Journal of the Australian Institute Agricultural Science and Technology)* **19** (3), 21–26.

Moran, J. B. (2009). *Business Management for Tropical Dairy Farmers.* Landlinks Press, Melbourne. <http://www.publish.csiro.au/nid/220/issue/5522.htm>

Moran, J. B. (2012). *Management of High Grade Dairy Cows in the Tropics.* CSIRO Publishing, Melbourne.

Moran. J. and McLean, D. (2001). *Heifer Rearing: A Guide to Rearing Dairy Replacement Heifers in Australia.* Bolworth Press, Melbourne.

Moran, J. B. and Tranter, B. (2004). *Reproductive Management of Small Holder Farmers in Vietnam.* F & N Vietnam Foods & DPI Kyabram, Hoh Chi Minh City, Vietnam.

Msanga, N. Y. and Bee, J. K. A. (2006). The performance of Friesian x Boran managed extensively under agropastoralism with indigenous Tanzanian Zebu. *Livestock Research for Rural Development* **18** (2). <http://www.lrrd.org/lrrd18/2/msan18020.htm>

NAHMS (National Animal Health Monitoring System) (1994). *Dairy Heifer Morbidity, Mortality, and Health Management, Focusing on Preweaned Heifers,* USDA, Center for Epidemiology and Animal Health, Fort Collins, Colorado, USA.

National Research Council (1989). *Nutrient Requirements of Dairy Cattle.* 6th edn. National Academy Press, Washington, DC, USA.

Nettisinghe, A. M., Udo, H. M. and Steenstra, F. A. (2004). Impact of an AI heifer calf rearing scheme on dairy stock development in the Western Province of Sri Lanka. *Asian-Australasian Journal of Animal Science* **17** (1), 18–26.

Ngategize, P. K. (1989). Economic evaluation of improved management for Zebu cattle in Northern Tanzania. *Agricultural Systems* **31**, 305–314.

Payne *et al.* (1967). Cited in <http://www.most.gov.mm/techuni/media/BioT_04042_pl_4.pdf>.

Quigley, J. D., Hammer, C. J., Russell, L. E. and Polo, J. (2005). Passive immunity in newborn calves. In *Calf and Heifer Rearing: Principles of Rearing the Modern Dairy Calf from Calf to Calving.* (Ed. P. C. Garnsworthy) pp. 135–157, Nottingham University Press, Nottingham, UK.

Roy, J. (1980). *The Calf.* 4th edn. Butterworths, Sydney.

Rufino, M. C, Herrero, M., van Wijk, M. T., Hemerik, L., de Ridder, N. and Gillier, K. E. (2009). Lifetime productivity of dairy cows in smallholder farming systems of the Central highlands of Kenya. *Animal* **3**, 1044–1056.

STOAS (1999). *Reproduction in Dairy Cattle.* Book 1. STOAS, Wageningen, The Netherlands.

Suzuki, K. (2005). Investigation into the constraints to dairy cattle health and production in north Vietnam. Ph.D Thesis. Royal Veterinary College, University of London, UK.

Tiberghien, D. (2009). *Introduction to Good Calf Rearing Practices.* Presented at 'The dissemination workshop of the Vietnam Belgian Dairy Project, Part 2, Innovative dairy farming,' November 2009, Hanoi, Vietnam.

Tiwari, R., Sharma, M. C. and Singh, B. P. (2007). Buffalo calf health care in commercial dairy farms: a field study in Uttar Pradesh (India). *Livestock Research for Rural Development* **19** (3). <http://www.lrrd.org/lrrd19/3/tiwa19038.htm>.

Vaccaro, L. (1990). Survival of European dairy breeds and their crosses with Zebus in the tropics. *Animal Breeding Abstracts* **58**, 475–493.

Vaccaro, L. D. and Vaccaro, R. (1981). Losses up to first calving in Brown Swiss x Zebu and Holstein Friesian x Zebu heifers in an intensive system of milk production in the tropics. *Tropical Animal Production* **6** (4) 308–317.

Webster, J. (1984). *Calf Husbandry, Health and Welfare.* Granada, Sydney.

Wymann, M. N. (2005). Calf mortality and parasitism in periurban livestock production in Mali. Ph.D thesis. University of Basel, Germany.

Index